ACCLAIM FOR *A Geography of Time*

"An elegant gem ... Levine takes us behind the lens of the sensitive observer's eye to make us aware of the psychology of time as perhaps the greatest of human inventions. He combines brilliant observations, original field experiments, and wide-ranging scholarship to generate an original view of how subjective time and human functioning mesh or collide. *A Geography of Time* is a worthwhile detour: take it and value its lessons well."
— Philip Zimbardo, author of *Psychology and Life*

"What a timely book! An empirically trained social psychologist casts an informed eye across the cross-cultural literature on how people in various parts of the globe structure their 24 hours each day. The 'silent language' of time is articulated in this pacey, humorous assessment of how this basic dimension of our lives affects us all. Scholarly but fun, informative but colorful. Take time to read this book."
— Michael Harris Bond, Chinese University of Hong Kong

"Our treatment of time turns out to be a masterful key that opens a fantastic array of doors into numerous intellectual, social, cultural, and many other-worldly areas."
— Amitai Etzioni, author of *The New Golden Rule*

"Anyone who picks up this book believing that time is simply something that is measured by that little gadget on your wrist is in for a major revelation and a mind-expanding experience (as well as a good 'time.') Levine is to be congratulated; truly, an excellent piece of work."
— Elliott Aronson, University of California at Santa Cruz

"Unique, wry, and readable, this well-documented book is recommended for social psychology collections and public libraries for sophisticated readers with the time to spare."
— *Library Journal*

"Levine shows with grace, wit, and scholarship how culture-bound our sense of time really is. *A Geography of Time* has altered (for the better) my own attitude toward time. This book should make a major contribution to breaking the shackles of time pressure that bind us all."
— Ralph Keyes, author of *Timelock*

"Packed with interesting observations and information."
— Anthony Storr, *Washington Times*

ABOUT THE AUTHOR

Robert Levine, Ph.D., is former Associate Dean, College of Science and Mathematics, and Professor of Psychology at California State University, Fresno, where he has received many awards for his teaching and research. He has been a visiting professor at Universidade Federal Fluminense in Niteroi, Brazil, at Sapporo Medical University in Japan, and at Stockholm University in Sweden. He has published articles in *Psychology Today, Discover, The New York Times,* and *American Scientist,* and has appeared on ABC's *World News Tonight, Dateline,* NBC, CNN, The Discovery Channel, and *All Things Considered.* He is also the editor, with Aroldo Rodrigues, of *Reflections On 100 Years Of Experimental Social Psychology.* His latest book, *The Power of Persuasion,* is also published by Oneworld.

A GEOGRAPHY OF TIME

The Temporal Misadventures of a Social Psychologist,
or
How Every Culture Keeps Time Just a Little Bit Differently

Robert Levine

ONEWORLD

A Oneworld Book

Published by Oneworld Publications 2006
Reprinted 2012, 2016, 2018

First published in the United States of America by Basic Books,
a member of the Perseus Books Group

ISBN-13: 978-1-85168-465-6
ISBN-10: 1-85168-465-4

Cover design by Mungo Designs
Printed and bound in Great Britain by Clays Ltd, Elcograf S.p.A.

Oneworld Publications
10 Bloomsbury Street
London WC1B 3SR
England

Stay up to date with the latest books,
special offers, and exclusive content from
Oneworld with our newsletter

Sign up on our website
oneworld-publications.com

MIX
Paper from
responsible sources
FSC® C018072

For Trudi, Andy and, of course, Mr. Zach

CONTENTS

PART III

Changing Pace

ACKNOWLEDGMENTS

I have been surrounded by so many helpful and knowledgeable students, colleagues, and friends that it is difficult to know where my ideas begin and where theirs leave off. Let me single out a few of these people for special thanks. I am indebted to the following for generously providing insights and stories used in this book: Neil Altman, Stephen Buggie, Kris Eyssell, Alex Gonzalez, Eric Hickey, James Jones, the late William Kir-Stimon, Shirley Kirsten, Todd Martinez, Kuni Miyake, Salvatore Niyonzima, Harry Reis, Suguru Sato, Jean Traore, Fred Turk, and Jyoti Verma. Here at my university, Sergio Aguilar-Gaxiola, Jean Ritter, Aroldo Rodrigues, and Lynnette Zelezny have been indispensable sources of information and support. Among many other helpful colleagues, I would like to thank Rick Block, Richard Brislin (whose teachings were the inspiration for chapter 9), Edward Diener and Harry Triandis for their teachings about the subjects of time and/or culture, and their willingness to respond to my many requests for data and information. I cannot say enough about the constant support of Phil Zimbardo—he is not only social psychology's most inspiring teacher, but perhaps its greatest *mensch*. I thank Suguru Sato and Yoshio Sugiyama at Sapporo Medical University, Lars Nystedt and Anna and Hannes Eisler at Stockholm University, and the administration at Universidade Federal Fluminense in Brazil for making my residencies at their universities such successful experiences; these adventures are at the heart of this book. I thank Ellen Wolff who began me on this project many years ago. My wife, Trudi Thom, consulted on many aspects of this project, and has always been there when I needed her. Alex Gonzalez, my colleague, boss, and fellow time traveler, has been a font of guidance and support during many incarnations of my research program. My colleague Connie Jones, who has read every word of this manuscript, has been an invaluable surrogate editor and all-around source of sustenance. Tom Breen, as always, has been just plain special.

I am indebted to the following people for their help in collecting and/or analyzing data in one or more cities in the United States or other countries: Timothy Baker, Laura Barton, Karen Bassoni, Stephen Buggie, Brigette Chua, Andy Chuang, Holly Clark, Lori Conover, John Evans, Kris Eyssell, David Hennessey, Kim Khoo, Robert Lautner, Marta Lee, Royce Lee, Andy Levine, Martin Lucia, Thom Ludwig, Allen Miller, Michiko Moriyama, Walter Murphy, Carlos Navarette, Julie Parravano, Karen Philbrick, Harry Reis, Aroldo Rodrigues, Michelle St. Peters, Anne Sluis, Kerry Sorenson, David Tan, Jyoti Verma, Karen Villerama, Sachiko Watanabe and Laurie West. Thanks to Philip Halpern for his technical input. Gary Brase, Karen Lynch, Todd Martinez, Kuni Miyake and Ara Norenzayan have played particularly pivotal roles in these studies, and I thank them all for their assistance and inspiration.

My agent, Kris Dahl, opened up doors that I didn't know existed. Elizabeth Kaplan came up with the winning title. Gail Winston, my editor at Basic Books, has been perfect. Her patient, positive, expert guidance has helped me hone a series of chapters into the semblance of a book. I have been fortunate to work with her.

PREFACE

TIME TALKS, WITH AN ACCENT

> Every culture has its own unique set of temporal fingerprints. To know a people is to know the time values they live by.
>
> JEREMY RIFKIN, *Time Wars*

Time has intrigued me for as long as I remember. Like most young Americans, I was initially taught that time is simply measured by a clock—in seconds and minutes, hours and days, months and years. But when I looked around at my elders, the numbers never seemed to add up the same way twice. Why was it, I wondered, that some adults appeared to be perpetually running out of daylight hours while others seemed to have all the time in the world? I thought of this second group of people—the ones who would go to the movies in the middle of the workday or take their families on six-month sabbaticals to the South Pacific—as temporal millionaires, and I vowed to become one of them.

When planning my career, I ignored my peers' unwavering concern with the amount of money a job would pay and tuned in instead to the temporal lifestyle it offered. To what extent would I be able to set my own pace: How much control would I have over my time? Could I take a bike ride during the day? Thoreau spoke to me when he observed, "To affect the quality of the day, that is the highest of arts." I chose a profession—that of a university professor—which offers the temporal mobility I sought. And to my good fortune, I encountered a specialty—social psychology—that has allowed me to pursue the very concept of time that fascinated me as a child.

I trace the beginning of my scientific journey to an experience early in my career. Until then, my research had been focused on what was at the time the hot topic in social psychology, attribution theory. I had confined my experiments to rather technical problems, such as how men and women differ in their explanations for success and failure, what conditions cause people to attribute their successes to external causes, and how self-confidence affects one's attributional style. You get the picture: these were significant issues within my own academic sphere, but I couldn't help noticing how my friends' eyes glazed over when I described my research.

My interest in these technical questions ended abruptly in the summer of 1976. I had just begun an appointment as a visiting professor of psychology at the Federal University in Niteroi, Brazil, a midsized city across the bay from Rio de Janeiro. I arrived anxious to observe at first hand just what characteristics of this alien environment would require the greatest readjustment from me. From my past travel experiences, I anticipated difficulties with such issues as the language, my privacy, and standards of cleanliness. But these turned out to be a piece of cake compared to the distress that Brazilians' ideas of time and punctuality were to cause me.

I was aware before arriving, of course, of the stereotype of the *amanhã* attitude of Brazilians (the Portuguese version of *a mañana*), whereby it is said that whenever it is conceivably possible the business of today is put off until tomorrow. I knew I'd need to slow down and to reduce my expectations of accomplishment. But I was a kid from Brooklyn, where one is taught at an early age to move fast or get out of the way. Years ago I had learned to survive life in the foreign culture of Fresno, California, a city where even laid-back Los Angelenos must learn to decelerate. Adjusting to the pace of life in Brazil, I figured, would call for no more than a bit of fine tuning. What I got instead was a dose of culture shock I wouldn't wish on a hijacker.

My lessons began soon after arriving. As I left home for my first day of teaching, I asked someone the time. It was 9:05 A.M., allowing me plenty of time to get to my 10 o'clock lecture. After what I judged to be half an hour, I glanced at a clock I was passing. It said 10:20! In panic, I broke for the classroom, followed by gentle calls of "*Alô, Professor*" and "*Tudo bem, professor?*" from unhurried students, many of whom, I later realized, were my own. I arrived breathless to find an empty room.

Frantically, I exited the room to ask a passerby the time. "Nine forty-five," came the answer. No, that couldn't be. I asked someone else. "Nine fifty-five." Another squinted down at his watch and called out proudly: "Exactly nine forty-three." The clock in a nearby office read 3:15. I had received my first two lessons: Brazilian timepieces are consistently inaccurate; and nobody seemed to mind but me.

My class was scheduled from ten until noon. Many students came late. Several arrived after 10:30. A few showed up closer to eleven. Two came after that. All of the latecomers wore the relaxed smiles I later came to enjoy. Each one greeted me, and although a few apologized briefly, none seemed terribly concerned about being late. They assumed that I understood.

That Brazilians would arrive late was no surprise, although it was certainly a new personal experience to watch students casually enter a classroom more than one hour late for a two-hour class. The real surprise came at noon that first day, when the class came to a close.

Back home in California, I never need to look at a clock to know when the class hour is ending. The shuffling of books is accompanied by strained expressions screaming: "I'm hungry / I'm thirsty / I've got to go to the bathroom / I'm going to suffocate if you keep us one more second." (The pain, I find, usually becomes unbearable at two minutes to the hour for undergraduates and at about five minutes to the hour for graduate students.)

But when noon arrived for my first Brazilian class, only a few students left right away. Others slowly drifted out during the next fifteen minutes, and some continued asking me questions long after that. When several remaining students kicked off their shoes at 12:30, I went into my own hungry/thirsty/bathroom/suffocation plea. (I could not, with any honesty, attribute their lingering to my superb teaching style. I had, in fact, just spent two hours lecturing on statistics in halting Portuguese. Forgive me, *meus pobres estudantes.*)

In the hope of understanding my students' behavior, I made an appointment for 11 A.M. the next morning with my new *chefe*, or department head. I arrived at her office on time. Neither she nor her secretary were there. In fact, I had to turn on the lights to read the magazines in the waiting room: a year-old copy of *Time* and a three-year-old copy of *Sports Illustrated.*

At 11:30 the secretary arrived, said *alô*, asked me if I wanted a *cafézinho* (the traditional Brazilian drink consisting of one-half thick coffee and one-half sugar, which, as best I can tell, gets everyone so wired that they no longer bother to move), and left. At 11:45 my new *chefe* arrived, also offered me a *cafézinho*, and also went off. Ten minutes later she returned, sat down at her desk, and began reading her mail. At 12:20, she finally called me into her office, casually apologized for making me wait, chatted for a few minutes and then excused herself to "run" to another appointment for which she was late. I learned later that this was no lie. It was her habit to make lots of appointments for the same time and to be late for all of them. She apparently liked appointments.

Later that day I had a meeting scheduled with several students from my class. When I got to my "office" two of them were already there and acting quite at home. They seemed undisturbed that I was a few minutes late and, in fact, were in no hurry to begin. One had kicked his feet up on my desk and was reading his *Sports Illustrated* (which, I noted, was only three months old).

Some fifteen minutes after the scheduled conclusion I stood up and explained that I had other appointments waiting. The students stayed put and asked pleasantly, "Who with?" When I listed the names of two of their associates, one fellow excitedly reported that he knew them both. He rushed to the door and escorted one of them from the waiting area—the other hadn't arrived yet—into my office. They all then proceeded to chit-chat and turn the pages of the *Sports Illustrated*. By the time his associate sauntered in, five minutes before the scheduled conclusion of our appointment, I was beginning to lose track of who was early and who was late—which, I was eventually to learn, was exactly the lesson that I should have been learning. For now, though, I was just plain confused.

My last appointment of the day was with the owner of an apartment I wanted to rent. This time I thought I could spot the little train coming. As soon as I arrived I asked his secretary how long I would have to wait. She said that her boss was running late. "How late?," I asked. "A half an hour, *mais ou menos*," she replied. Would I like a *cafézinho*? I declined and said I'd be back in twenty minutes. Upon my return, she said it would be a little while longer. I left again. When I came back ten minutes later, she told me her boss had gotten tired of waiting for me and had left for the day. When I began to snap out an angry message to give to her Sr. Landlord, the secretary explained that I'd left him no choice but to skip out on me. "Don't you understand, he's the owner and you're not. You're an arrogant man, Dr. Levine." That was the last time I tried to outmaneuver a Brazilian at the waiting game.

During my year in Brazil, I was repeatedly bewildered, frustrated, fascinated, and obsessed by the customs and ideas of social time that Brazilians sent my way. The reason that Brazilians' rules of punctuality so confused me, it soon become apparent, was that they are inseparably intertwined with cultural values. And when we enter the web of culture, answers come neither simply nor cleanly. Cultural beliefs are like the air we breathe, so taken for granted that they are rarely discussed or even articulated. But there is often a volatile reaction when these unwritten rules are violated. Unsuspecting outsiders like myself can walk into a cultural minefield.

No beliefs are more ingrained and subsequently hidden than those about time. Almost thirty years ago anthropologist Edward Hall labeled rules of social time the "silent language."[1] The world over, children simply pick up their society's conceptions of early and late; of waiting and rushing; of the past, the present, and the future. There is no dictionary to define these rules of time for them, or for strangers who stumble over the maddening incongruities between the time sense they bring with them and the one they face in a new land.

Brazil made it clear to me that time was talking. But understanding what it was saying was no simple matter. After several months of temporal blunders, I designed my first systematic experiments about time in an attempt to understand Brazilians' beliefs and rules about punctuality. This work, at first to my frustration but eventually my appreciation, raised more questions than it answered. What I found so intrigued me that I have spent most of the past two decades continuing to research both the psychology of time and the psychology of places. My research has evolved from studies of punctuality to those about the broader pace of life; further study has raised questions about the consequences the pace of life has for the physical and psychological well-being of people and their communities. This work has taken me through many of the cities of the United States and across much of the rest of the world. It has confirmed my earliest intuitions: that how people construe the time of their lives comprises a world of diversity. There are drastic differences on every level: from culture to culture, city to city, and from neighbor to neighbor. And most of all, I have learned, the time on the clock only begins to tell the story.

THE PSYCHOLOGY OF PLACE

As a social psychologist, I have come, on many levels, to appreciate the value of studying time in general, and the pace of life in particular. The discipline of social psychology casts a wide net. Unlike our colleagues in the fields of personality psychology and sociology—the first of whom tend to focus on the private, internal functioning of people, and the latter on their social groups—social psychologists are concerned with the give and take between individuals and the groups that guide their behavior. We study with no small arrogance what our founding father Kurt Lewin called the "life space," the sum total of the behavior of individuals as they exist in their environments: The whole nine yards.

The work reported in subsequent chapters begins with the assumption that places, like people, have their own personalities. I fully concur with sociologist Anselm Strauss that "the entire complex of urban life can be thought of as a person rather than a distinctive place, and the city can be endowed with a personality of its own."[2] Places are marked by their own cultures and sub-cultures, each with their unique temporal fingerprints.

I have pursued these fingerprints through studies of the time sense of geographical locales. My goal has been to systematically study how places differ in their pace of life, and how great these differences are. Classifying the social psychology of places is an inherently messy enterprise. My charge, I believe, is to reduce the noise as much as possible. The goal is not to discover invariant differences between places; it is to describe whatever differences do exist as carefully as possible. These studies are, in a sense, objective, empirical tests of the raw material of popular stereotypes.

To an empirical researcher, the enormity and diffuseness of the concept of pace of life can be problematic; it often leads down dauntingly murky paths. The topic sets off chains of associations and tangents about so many aspects of time—for example, the time of physics, biology, health, culture, personal relationships, music, art—that it sometimes creates a mass free association about experience itself. Asking about the experience of time is a little like the question "What is art?" Both themes probe so deeply and broadly into personal experience that they often evolve into questions such as "How should I live my life?," or its cousin, "What is the meaning of life?" These are interesting topics, to be sure, but a bit unwieldy for a researcher striving for methodological precision.

There are dangers in making generalizations about the characteristics of places, particularly when they are directed at the collective "personalities" of their people. The notion of applying a single set of characteristics to an entire population, or to any group for that matter, makes for sloppy thinking. The fact is that individuals in any setting differ vastly. Assigning global labels to the people of a particular city or country is overstereotyping; as such, it is potentially malicious.

But while it may be careless to overgeneralize about the people from a single place, it would be naive to deny the existence of significant, overall differences between places and cultures. Of course, many Italians resemble the time-focused stereotype of the Swiss more than they do Marcello Mastroianni (just ask the Milanese); some Brazilians are more driven than the average New Yorker. But there is evidence that, taken as a whole, the Swiss do tend to be more clock-conscious than the Italians and that the Cariocas of Rio are generally more laid back than New Yorkers. In any given situation, the most driven Type A may be more relaxed than the highest scoring Type B. All things being equal, however, the reverse is more likely to be the case.

The degree of variation in a culture may be viewed as a telling characteristic in itself. In Japan, for example, conformity is considered a virtue. There is a well-accepted Japanese saying: "The nail that sticks up is quickly hammered down." As a result, there is considerably greater public uniformity in Japan than within an individualistic culture such as the United States, where "The squeaky wheel gets the grease." The extent of cultural tightness represents a significant difference between these two countries.

My studies compare the pace of life of different places, and range from early experiments comparing Brazil and the United States to recent ones comparing 31 different countries. One goal of this research has been to rank the paces of different cities and countries—to create a sort of social psychologist's places-rated guide to the fastest and slowest places to live. These lists follow an old American tradition. As early as the seventeenth century, promoters of Maryland were trying to persuade colonists to choose their state over Virginia by compiling statistics showing heavier turkeys, more deer, and fewer deaths from foul summer diseases and Indian massacres—all theirs for settling on the northern shores of Chesapeake Bay. Today's ratings of places provoke more debate than ever. As a *Time* magazine reporter put it, "Whether the subject is the beefiest burger or the biggest corporation, Americans have a penchant for making lists of the best and the worst, then arguing about the results. No rankings have inspired more disagreement than those about home sweet home."[3]

But whereas most places-rated studies have relied on statistics concerning objective living conditions (housing, health care, crime, transportation, education, the arts, recreational facilities, jobs), my research has explored the social-psychological quality of peoples' lives. How accurate is our stereotype of laid-back Southern Californians? Of Type A New Yorkers? How does the pace of life in Japan compare to that in Indonesia? In Syria versus Brazil? In which cities are people more likely to take the time to help a stranger? These are some of the questions I have asked along with my colleagues.

My own overriding goal in rating places has been more than deciding the "best" or "worst" cities. Rather, it has been to understand the consequences the pace of life holds for the quality of peoples' lives. Are people in slower places happier than their Type A counterparts? Are they healthier? Do they invest more time in their social responsibilities? My colleagues and I have examined the consequences of the pace of life on several levels, which range from the economic and social characteristics of cities to helping behavior to mortality rates from coronary heart disease.

Differences in the pace of life turn out, as we shall see, to have far-reaching consequences. This shouldn't be surprising. After all, the pace of our lives governs our experience of the passage of time. And how we move through time is, ultimately, the way we live our lives. As J. T. Fraser, the founder of the International Society for the Study of Time, wrote, "Tell me what to think of time, and I shall know what to think of you."

This book is *not* intended as yet another statement about the "overworked American," or the "time crunch," or "urgency addiction"—although I will touch on these subjects. And it is certainly not designed to be a time-management or self-help book—though here again I will try to offer a few suggestions that have come out of my work. Many fine books already exist on these subjects.[4] My interest is broader. In *A Geography of Time* I seek to understand the richness and complexity of views about time and the pace of life among cultures and cities and people around the world. Since time is the very cornerstone of social life, the study of a people's temporal constructions offers a precious window into the psyche of culture, including our own.

In researching other places I have learned as much about my own culture as I have about those of others. The social scientist's exploration of other people and places is not very different from any other type of travel writing in this respect. Both, if they are any good, should ultimately shed new light on life closer to home. As the writer Russell Banks once said at a symposium on the literature of travel:

> All travel writing that's of lasting interest—writing that we continue to read, writing that is written by writers as travelers, not travelers as writers—is really written to make a point about home. The essential purpose of diligent and brave observation of the other is to clarify the nature and the limits of the self, which leads one to conclude that the best travel writers are people who, needing that clarification, are at bottom unsure of the nature and limits of home and their relation to it. They move out of the house. So that like Hawthorne's Wakefield, they can look back and see what's true there.[5]

If I have done my job well, this book will set a clearer focus on the pace of our own lives as well as that of others. How do we use our time? What is this use doing to our cities? To our relationships? To our own bodies and psyches? Are there decisions we have made without consciously choosing them? Alternative tempos that we might prefer? Perhaps we can be led, like Hawthorne's Wakefield, to "look back and see what's true there," and, in our own ways, to achieve temporal prosperity.

A
GEOGRAPHY
OF
TIME

PART I

SOCIAL TIME

The Heartbeat of Culture

ONE

TEMPO

The Speed of Life

The question of tempo . . . depends not only on the
factors of personal taste and skill but to some extent
upon the individual instrument and the room or hall
involved in the performance.

WILLARD PALMER,
Chopin: An Introduction to His Piano Works

The pace of life is the flow or movement of time that people
experience. It is characterized by rhythms (what is the pattern of work time to down time? is there a regularity to social activities?), by sequences (is it work before play or the other
way around?), and by synchronies (to what extent are people and
their activities attuned to one another?). But first and foremost,
the pace of life is a matter of tempo.

The term "tempo" is borrowed from music theory, where it
refers to the rate or speed at which a piece is performed. Musical
tempo, like the time of personal experience, is extremely subjective. At the top of virtually every classical score, the composer in-
serts a nonquantitative tempo mark—*largo* or *adagio* to suggest a
slow tempo, *allegro* or *presto* for fast tempos, *accelerando* or *ritar-*

dando for changing tempos. There is even a directive called *tempo rubato*—literally translated as "stolen time"—which calls for a give-and-take in tempo between the two hands. But unless the composer specifies a metronome setting (which most classical composers did not or could not do, as the metronome was not marketed until 1816), the precise metric translation of the notation is open to widely varying interpretation. Depending on the speed at which the performer sets the metronome, Chopin's *Minute Waltz* may take up to two minutes to play.

The same is true for human time. We may play the same notes in the same sequence, but there is always that question of tempo. It depends upon the person, the task and the setting. One student may stay up all night to learn the same material that a gifted friend absorbs in an evening. The novelist might wait patiently for his next image, while his fellow writer at the newspaper races from deadline to deadline. Given an hour to spare with their child, one parent uses it to read aloud; another teams up in a demanding video game. My college student cousin travels around Europe for two months while his businessman father hurries across the same route in two weeks.

The speed may be measured over brief and immediate periods of time, as when one experiences rapidly oncoming traffic or an upcoming deadline, or over longer, more sustained intervals, such as when we speak of the accelerating tempo of twentieth-century living. Alvin Toffler, for example, in his popular book *Future Shock*, addresses the subject of tempo when he speaks of the psychic disruption that is caused by too much change in too short a time. The trauma is not caused by the shock of change per se, but by the rate of change. Whether considered over the short or the long term, and no matter how it is measured, there are vast cultural, historical, and individual differences in the tempo of life.

Time Signatures Around the World

> The further East I travel the sloppier the perception
> of time becomes. It irritates me in Poland and drives
> me gibbering in the USSR.
>
> ANONYMOUS BRITISH TRAVELER

Adjusting to an alien tempo can pose as many difficulties as learn-ing a foreign language. In one particularly telling study of the roots of culture shock, sociologists James Spradley and Mark Phillips asked a group of returning Peace Corps volunteers to rank 33 items as to the amount of cultural adjustment each had re-quired of them. The list included a wide range of items familiar to travel paranoids, such as "the type of food eaten," "personal clean-liness of most people," "the number of people of your own race" and "the general standard of living." But aside from mastering the foreign language, the two greatest difficulties for the volunteers concerned social time: "the general pace of life," followed by one of its most significant components, "how punctual most people are."[1]

   Neil Altman was one of these temporally disoriented Peace Corps volunteers. Altman, who is now a clinical psychologist in New York City, served a term as an agricultural consultant in a vil-lage in the South of India. "When we first got to India," he recalls, "I used to go to the local horticulture office to get seeds and the like. I'd go into the office of the head guy to request what I wanted, but would find six or eight people sitting around his desk, each person with some business to transact, presumably. I would impatiently state my purpose: 'Good morning, Mr. Khan, could I get some tomato seeds, please?' 'Good morning, Volunteer *sahib*, won't you join us for some tea?' So I would have no choice but to sit down and wait while some servant ran out to get me tea. Then Mr. Khan would inquire about my wife, etc., and then all the as-sembled people would have a million questions about my life, about America, etc., etc., etc. It would be hard to know how to ask for my tomato seeds again without being rude. Eventually, after an hour or two I would decide to risk being rude anyway. I would get my seeds and be on my way, noting that none of the people sitting around the desk had gotten any of their business taken care of."[2]

My own travels to the third world have led to the same confrontations with tempo. Sometimes it seems life in these countries is one long wait: for buses and trains, for entry and exit visas, for dinner, for toilets. Once, when trying to get to the train station in New Delhi, I waited 45 minutes for a bus so crowded that I had to hang on an extra two stops until I could force my way off. From there, I walked back to the station, where I waited nearly another hour to buy my train ticket. When I finally got to the window, the cashier greeted me with the traditional "*Namasté*" and immediately flipped up a sign that read "Closed for Lunch" (in English, I might add). With my blood pressure headed for Kashmir, I turned around to gather support for my case. But all of my compatriots were already sitting on the floor, with their blankets spread out, eating picnic lunches. "What can I do?" I asked a couple next to me. "You can join us for lunch," they answered. After several false starts to nowhere, I finally did.

When the ticket window reopened, I found my position in line had been taken by a family of six. They offered me peanuts and blessed me in Hindi. When I asked them to give me my place back, the eldest male smiled politely and mumbled something that I swear sounded like "When Shiva flies to Miami Beach." When I finally got to the ticket window, I was told my train was sold out. And all this work was for a train that was not going to leave for three more days. I eventually did get a ticket (oh, the miracles that a little *baksheesh* [bribery] can accomplish). But even with a ticket, I was told to come to the station an hour early, only to find I had to push through several waves of crowds to ask someone to get out of my reserved seat. Needless to say, the train left late and arrived even later, none of which mattered, because the gentleman I was scheduled to meet at the station was even later than me.

There is an inscription on the narrow-gauge Darjeeling Himalayan Express that reads: "'slow' is spelled with four letters; So is 'life.' 'Speed' is spelled with five letters; So is 'death.'" Really.

Intercultural struggles over tempo are found all over the world. My colleague Alan Button, for example, tells how he was once late for an appointment while traveling in Russia. His guide began shouting to their cab driver a Russian phrase (*Pah yeh kaly*) meaning "Get there yesterday" or literally, "Let's went." His guide advised him that the literal translations of words like "hurry" and "rush" simply do not carry the urgency in Russian that they do in English. If he had merely ordered the driver to "Get there as soon as you can," Button was told, he would have arrived even later than he did. As it turns out, he arrived very late, but still found that he was some twenty minutes earlier than the fellow he was scheduled to meet.

The literature is filled with accounts of rushing, time-is-money travelers whose racing leaves the baffled residents of slower worlds running for cover. During my year in Brazil, it seemed as if I heard no more frequent words from my laid-back hosts than their pleading advice: "*Calma, Bobby, calma.*" No matter how hard I tried to slow down, there almost always seemed to come the breathless "*Calma, por favor*"—sometimes as an appeal, other times offered with head-shaking pity. And I was simply moving at the tempo of a college professor from Fresno—hardly America's prototype of hurriedness.

James Jones, a fellow social psychologist from the University of Rhode Island, had a similar experience when he was living in the West Indian nation of Trinidad several years ago. He had traveled to Trinidad on a Guggenheim Fellowship to study its people's humor. But what he learned more than anything was that he was always seriously out of step. Latecomers to appointments, he reports, would greet his impatience with comments like: "Eh mon, what's your hurry, nuh? De sea ain goin' no place. Relax mon, a'm comin' to yuh just now." "So," as Jones put it, "I wait." Perhaps the most remarkable similarity in Jones's and my experiences was the profound results they had for our careers. Although we both had limited success in achieving the original goals of our projects—his to study humor in Trinidad and mine to study social perception in

Brazil—these interests soon receded into the background. The more compelling puzzle, to both the traveler and the social psychologist in each of us, was the richness of the social time we encountered and our confusion with it. As a result, the study of time has become the focus of each of our research programs. Jones has gone on to become an international authority on the psychology of time perspective, and I have remained obsessed with studying the pace of life.

ELEMENTS OF TEMPO

What characteristics of places and cultures makes them faster or slower? To answer this question, my own research group has recently completed a series of studies comparing the pace of life in 31 different countries from throughout the world. The results of these experiments, coupled with research findings from other social scientists, establish several factors that are critical in the establishment of tempo norms.

Let me briefly describe how my studies (to which I will return in more detail in a later chapter) were conducted. In each country, we went into one or more of the major cities in order to measure three indicators of the tempo of life.[3] (For simplicity, Hong Kong is referred to here as a country despite its present colonial status.)[4] First, we measured the average walking speed of randomly selected pedestrians over a distance of 60 feet. The measurements were made on clear summer days during main business hours, usually during the morning rush, in at least two locations on main downtown streets. Locations were chosen that were flat, unobstructed, had broad sidewalks, and were sufficiently uncrowded that the pedestrians could potentially walk at their own preferred maximum speed. In order to control for the effects of socializing, only pedestrians walking alone were used. Neither subjects with clear physical handicaps nor those who appeared to be window-shopping were timed. A minimum of 35 walkers of each sex were clocked in each city.

The second experiment focused on an example of speed in the workplace: the time it took postal clerks to fulfill a standard request for stamps. In each city, we presented clerks with a note in the local language requesting a common stamp—the now standard 32-center in the United States, for example. They were also handed paper money—the equivalent of a $5 bill. We measured the elapsed time between the passing of the note and the completion of the request.

Third, as an estimate of a city's interest in clock time, we observed the accuracy of 15 randomly selected bank clocks in main downtown areas in each city. Times on the 15 clocks were compared to those reported by the phone company.

The three scores for each country were then statistically combined into an overall pace-of-life score.

From these experiments and the research of others, one can determine five principal factors that determine the tempo of cultures around the world. People are prone to move faster in places with vital economies, a high degree of industrialization, larger populations, cooler climates, and a cultural orientation toward individualism.

Economic Well-Being
The healthier a place's economy, the faster its tempo.

As a city grows larger, the value of its inhabitants' time increases with the city's increasing wage rate and cost of living, so that economizing on time becomes more urgent, and life becomes more hurried and harried.

IRVING HOCH[5]

The number one determinant of a place's tempo is economics. Without question, the strongest and most consistent finding in our experiments is that places with vital economies tend to have faster tempos. The fastest people we found were in the wealthier North American, Northern European, and Asian nations. The slowest were in third-world countries, particularly those in South and Central America and the Middle East. (See chapter 6 for more detailed results.)

Faster overall tempos are highly related to a country's economic well-being on every level: to the economic health of the country as a whole (as measured by gross domestic product per capita); to the economic well-being actually experienced by the average citizen (as measured by purchasing power parity, which is an estimate of how much the average income earned in a country is capable of purchasing); and to how well people are able to fulfill their minimum needs (measured by average caloric intake).[6] People from richer and poorer nations do, in fact, march to different drummers.

We can speculate about the direction of causality between the tempo of life and economic conditions. Most likely, the arrow points both ways. Places with active economies put greater value on time, and places that value time will be more likely to have active economies. Economic variables and the tempo tend to be mutually reinforcing; they come in a package.

We don't need to travel to other countries to see the connection between economics and tempo. Some of the most telling evidence for the economic explanation appears in subcultures within countries. In the United States, for example, many economically impoverished minority groups make a point of distinguishing their own shared temporal norms from those of the prevailing Anglo-American majority. American Indians like to speak of "living on Indian time." Mexican-Americans differentiate between *hora inglesa*—which refers to the actual time on the clock—and *hora mexicana*—which treats the time on the clock considerably more casually.

African-Americans often distinguish their own culture's sense of time—what they sometimes refer to by the no longer fashionable term "colored people's time" (CPT)—from the majority standard of "white people's time." Jules Henry, an anthropologist, spent more than a year during the 1960's conducting interviews with mostly poor African-American families living in a St. Louis housing development. One of the strongest distinctions his inter-viewees made between their own lives and those of the surround-ing Anglo community concerned their self-described CPT. "According to C.P. time," Henry explains, "a scheduled event may occur at any moment over a wide spread of hours—or perhaps not at all." Henry's interviewees were quick to point out how sharply this contrasted with the highly organized, precisely scheduled world of white people.[7]

The sociologist John Horton applies a more contemporary slant to CPT, using it to refer to "cool people's time." The term "cool people" refers to the "sporadically unemployed young Black street corner population." Horton spent two years interviewing many of these street people. "Characteristically," he reports,

> the street person gets up late, hits the street in the late morn-ing or early afternoon, and works his way to the set. This is a place for relaxed social activity. Hanging on the set with the boys is the major way of passing time and waiting until some necessary or desirable action occurs . . . On the set yesterday merges into today, and tomorrow is an emptiness to be filled in through the pursuit of bread and excitement[8].

The prevailing tempo, in other words, is very slow. As Horton makes clear, however, the street people are adept at speeding up their tempos when the situation calls for it. The street dude, ac-cording to Horton, is on time by the standard clock whenever he cares to be and is not on time when he doesn't want to be. Most of-ten, the latter is the case. Time for the cool person is "dead" when resources are low—such as when money is tight, or when he's in jail. But time is "alive" whenever and wherever there is "action." The tempo is slow early in the week, when money is tight, but ac-celerates exponentially on Friday and Saturday nights.[9]

The Degree of Industrialization

The more developed the country, the less free time per day.

What kind of rule is this? The more timesaving machinery there is, the more pressed a person is for time.

SEBASTIAN DE GRAZIA,
Of Time, Work, and Leisure

We should not be surprised that the wealthier places in our experiments have faster norms. Economic vitality is closely tied to industrialization. Historically, in fact, the single most crucial watershed event in the acceleration of the tempo of the Western world was the Industrial Revolution.

It is one of the great ironies of modern times that, with all of our time-saving creations, people have less time to themselves than ever before. Life in the Middle Ages is usually portrayed as bleak and dreary, but one commodity people had more of than their successors was leisure time. Until the Industrial Revolution, in fact, most evidence suggests that people showed little inclination to work. In Europe through the Middle Ages, the average number of holidays per year was around 115 days. It is interesting to note that still today, poorer countries take more holidays, on the average, than richer ones.

It has often been the very creations intended to save time that have been most responsible for increasing the workload. Recent research indicates that farm wives in the 1920's, who were without electricity, spent significantly less time at housework than did suburban women, with all their modern machinery, in the latter half of the century. One reason for this is that almost every technical advance seems to be accompanied by a rise in expectations. For example, when cheap window glass was introduced in Holland at the end of the seventeenth century it became impossible to ignore the dirt that accumulated indoors. Today's vacuum cleaners and other products have raised peoples' cleanliness standards even higher; in so doing, they demand that people invest the time needed to propel these products against the suddenly defeatable household grit and bacteria.[10] So much for better living through Westinghouse.

It is telling to observe how modern conveniences have affected the way people use their time. A study by anthropologist Allen Johnson, for example, compared the use of time among the Machiguenga Indians to that of workers living in France. The French workers, he found, spend more time at work and consuming things (eating, reading, watching television), but have considerably less free time than the Machiguenga workers. These differences are true for both men and women. French men spend four times as many hours consuming the fruits of their labor, but pay a stiff price for these goodies: They have four hours less free time per day than their Machiguenga counterparts. Perhaps most tellingly, Johnson found that the conveniences of modern living extract an extremely high toll in the time required for their maintenance. The Machiguenga give three to four times more of their production time at home to manufacturing (for example, baskets and cloth) than they do to maintenance work (doing the laundry, cleaning, making repairs). The French pattern is almost exactly the reverse. In the end, as the anthropologist Marvin Harris observed, modern appliances are "labor-saving devices that don't save work."

Johnson, borrowing from recent economic theory, argues that industrialization produces an evolutionary progression from a "time surplus" to a "time affluence" to a "time famine" society, which is how he characterizes most developed countries. The ultimate effect, Johnson argues, is on the tempo of people's lives:

> As a result of producing and consuming more, we are experiencing an increasing scarcity of time. This works in the following way. Increasing efficiency in production means that each individual must produce more goods per hour; increased productivity means . . . that to keep the system going we must consume more goods. Free time gets converted into consumption time because time spent neither producing nor consuming comes increasingly to be viewed as wasted . . .

The increase in the value of time (its increasing scarcity) is felt subjectively as an increase in tempo or pace. We are always in danger of being slow on the production line or late to work; and in our leisure we are always in danger of wasting time.[11]

At the slow extreme of the tempo continuum are the Stone Age economics of so-called primitive agricultural and hunting-gathering societies. The Kapauku of Papua, for example, don't believe in working two consecutive days. The !Kung Bushmen work two-and-a-half days per week, typically six hours per day. In the Sandwich Islands, men work only four hours per day.[12]

On the average, studies show, women in less advanced economies work an average of 15 to 20 hours per week, and men put in about 15 hours. The shift to plow cultivation, which requires feeding and caring for draft animals, pushes the work week of men to 25 to 30 hours. It requires one day for a Dobe woman in Australia to gather enough food to feed her family for three days. The rest of the time is her own—to visit, entertain, work on her embroidery, or, as is often the case, to do nothing at all.

There are some underdeveloped cultures where the clock seems to stand still, if it exists at all. Edward Hall, an anthropologist, relates the story of an Afghani man in Kabul who could not locate a brother with whom he had an appointment. An investigation by a member of the American embassy eventually revealed the root of the problem: The two brothers had agreed to meet in Kabul, but had neglected to specify what year.[13] What often surprises clock-watching Anglo-Europeans most about this story is to learn just how many people in the world fail to see the humor in Hall's story; most are quite understanding and sympathetic toward the miscommunication.

But it would be a gross generalization to conclude that industrialization and tempo are one and the same. Sometimes the tempos of third-world cultures can be strikingly different, even between seemingly similar neighbors. The anthropologist Paul Bohannan, for example, has researched tribal greeting styles.[14] In one study, he compared the Tiv of Nigeria to their neighbors, the Hausa. The Tiv, he found, are fast people. They waste little time with perfunctory rituals such as greetings. They like to get their hellos out of the way quickly and get right down to business. Living right next to these third-world Type A's are their neighbors, the Hausas, who would not think of depriving a greeting of its rightful duration. Bohannan tells of having once observed an English anthropologist and a Hausa string out their hellos for 20 minutes. They both seemed to enjoy the ritual, the intricacies of which they had been practicing and perfecting for many years.

Rules governing how soon a greeting should begin may also vary. Sushila Niles, currently a psychology instructor at Northern Territory University in Darwin, Australia, tells about an unpleasant encounter with a government official during her stay as a teacher in an African country. After being sent in by the man's secretary, Niles found him in conversation with someone else. "I stood aside politely," she recalls. "Suddenly he turned to me and said 'What madam, no greeting?' I had breached all conventions of social interaction by not greeting him the moment I stepped into his office. I said that I had been brought up to believe that interrupting was rude. But he was not mollified."[15]

Stephen Buggie, a professor of psychology at Presbyterian College in South Carolina, spent three years teaching in Zambia and nine years in Malawi. "In Zambia," he recalls, "the tempo of life is generally slow, with casual regard toward punctuality and time. But walking speed in downtown Lusaka (the capital and largest city) is fast, as an individual deterrent against rampant pickpocketing. Neighboring Malawi is very different. Meetings there start more promptly than in Zambia. Malawi's Life President, Kamuzu Banda, practiced medicine in Scotland for 30 years before entering politics back home. He rules the country absolutely and is a stickler for punctuality. Back in the 1970's he made it illegal for public clocks to display inaccurate time. Broken clocks were supposed to be removed or covered with a shroud."[16]

Population Size

Bigger cities have faster tempos.

After economic well-being, the single strongest predictor of differences in the tempo of places is population size. Studies have shown over and over again that, on the whole, people in bigger cities move faster than their counterparts from smaller places.[17]

In one of the earliest studies of this type, Herbert Wright, as part of his classic "City-Town" project, observed the behavior of children in typical city supermarkets and in small-town grocery stores. One of the strongest differences between the two environments turned out to be walking speed. The average city child walked nearly twice as fast through the supermarket as the town child did through the smaller grocery. The town children spent three times as much time interacting with clerks and other shoppers. They also spent significantly more time physically touching objects in the market.[18]

Australian psychologist Paul Amato found comparable differences on the other side of the world, in New Guinea. In an interesting series of experiments, Amato observed pedestrian walking speed, the speed with which change was given in European shops, and the elapsed time of betel-nut transactions in open marketplaces in a large city (Port Moresby) and two rural towns (Wewak and Mount Hagen). The urban locale clocked in with faster speeds on the walking measure and betel-nut transactions. There were no urban-rural differences on the change measure—tellingly, no one in any of the locales in New Guinea seemed at all interested in this sort of activity.[19]

The definitive treatise on the association between walking speed and population size comes from a series of international studies by psychologist Marc Bornstein and his colleagues. In their first group of experiments, Bornstein's team observed walking speeds in main downtown locations in a total of 25 cities spread across Czechoslovakia, France, Germany, Greece, Israel, and the United States. They found an astonishingly high relationship between population size and walking speed across this heterogeneous collection of cities. (In statistical terms, they found a correlation of r = .91 between population size and walking speed, with 1.00 being the highest correlation possible; in other words, an almost perfect relationship.) [20]

When strong mathematical relationships occur in cross-cultural studies of this type, they beg for replication. Answering this challenge, Bornstein conducted a second series of studies. He applied the conditions of his earlier investigation to a new sample of cities and towns in Ireland, Scotland, and the United States. Once again, he found that there was an extremely strong correlation between population size and walking speed (r = .88). Bornstein argues that "a highly predictable relationship seems to exist between the pace of life that characterizes a locale and the size of its population."[21] Given Bornstein's results—one does not often discover correlations of this magnitude in the inherently noisy science of social psychology—it is difficult to argue with his conclusion.

Climate
Hotter places are slower.

There is also considerable validity to the old stereotype about life being slower in warmer places.[22] All of the slowest nations in our 31-country study—Mexico, Brazil, and Indonesia were the slowest of all—have tropical climates. These are the sort of places that people from the fastest countries—Switzerland, Ireland, Germany—look toward for their winter vacations. Looking at the 31 countries as a whole, we found a strong relationship between the climate (as measured by average maximum temperatures)[23] of cities and how slow they were on our measures.

Some people believe that the slow tempo of warm places has an ergonomic explanation-that it results from a general lack of energy. Certainly, anyone who has been through a heat wave knows that high temperatures can wear one down. Others hypothesize that the slowness has an evolutionary/economic sensibility. They argue that people in warmer places don't need to work as hard. They require fewer and less costly belongings—fewer clothes, simpler homes, so why bother to rush? Then there are people who believe that warmer climates simply encourage taking time to enjoy life. Whatever the explanation, it is clear that hotter places are much more likely to have slower tempos.

Cultural Values
Individualistic cultures move faster than those that emphasize collectivism.

A culture's basic value system is also reflected in its norms about tempo. Probably the strongest cultural differences concern what is known as individualism versus collectivism: whether the basic cultural orientation is toward the individual and the nuclear family or to a larger collective. The United States is a classic individualistic culture. Traditional Asia, on the other hand, tends to focus on the collective. In Pakistan and India, for example, many people share large homes with their extended families—something on the order of individual apartments with shared kitchen facilities. In Tibet and Nepal, families live together, and it is common for brothers to share the same wife—an economically convenient arrangement for Sherpas (porters) who spend most of their lives away from home. In some collectivistic cultures, the sense of family extends toward the entire village, or even the national "tribe." Many cross-cultural psychologists believe that the individualism-collectivism continuum is, in fact, the single most significant characteristic of the social patterns of a culture.

Harry Triandis, a social psychologist at the University of Illinois who is considered the foremost expert on the topic of individualism-collectivism, has found that individualistic cultures, compared to collectivist ones, put more emphasis on achievement than on affiliation.[24] The focus on achievement usually leads to a time-is-money mindset, which in turn results in an urgency to make every moment count. In cultures where social relationships take precedence, however, there is a more relaxed attitude toward time. Collectivist cultures, then, should be characterized by slower tempos. We tested this prediction in our 31-country study by comparing each country's individualism-collectivism scores[25] to their times on our three experiments. Our results confirmed the hypothesis: greater individualism was highly related to faster tempos.

A focus on people, as we shall see in subsequent chapters, is often at odds with a tempo dictated by schedules and the time on the clock. In some collectivist cultures, in fact, time urgency is not only deemphasized but treated with downright hostility. The anthropologist Pierre Bourdieu, for example, has studied the Kabyle people, a collectivist society in Algeria. The Kabyle, he found, want nothing to do with speed. They despise any semblance of haste in their social affairs, regarding it as a "lack of decorum combined with diabolical ambition." The clock is referred to as "the devil's mill"![26]

THE BEAT OF YOUR OWN DRUM

Time travels in divers paces with divers persons
WILLIAM SHAKESPEARE, *As You Like It*

This book focuses on the pace of life as it differs between cultures and places. Obviously, however, there are also vast differences in tempo between individuals within the same culture, as well as between those living in the same town or city. Neighbors may vary in both their personal preferences and in the tempo of life they actually experience.

Most of the attention to individual differences has centered on the concept of time urgency—the struggle to achieve as much as possible in the shortest period of time. Time urgency is one of the defining components of the Type A behavior pattern. Meyer Friedman and Ray Rosenman described the coronary-prone personality as being impatient, having a tendency to walk quickly, eat quickly, do two things at once, and to take pride in always being punctual.[27] The most widely used test of Type A behavior, the Jenkins Activity Survey, measures these characteristics with a "Speed and Impatience" scale.[28] Several more recent Type A scales have been developed, with labels such as "Time Urgency," "Perpetual Activation,"[29] and "Timelock."[30] All of these tests find extensive individual differences in the degree to which people are concerned with making every moment count.

But we need to be careful not to overgeneralize about "fast" and "slow" people. As is the case for cultures, each individual's pace may vary sharply according to the time, the place, and what they are doing. If you would like an accurate gauge of your own tendency toward time urgency, it is important to look into a wide range of behaviors. You might begin by thinking about yourself in these ten areas:

- Concern with clock time: Compared to most people, are you particularly aware of the time on the clock? Do you, for example, frequently glance at your watch? Or, on the other hand, are you the sort of person who sometimes forgets the time or even what day of the week it is?

- Speech patterns: How rushed is your speech? Do you tend to speak faster than other people? When someone takes too long to get to the point while speaking, do you often feel like hurrying them along? Are you a person who accepts interruptions?

- Eating habits: How rushed is your eating behavior? Are you often the first person finished eating at the table? Do you take time to eat three meals a day in a slow and relaxed manner?

- Walking speed: Do you walk faster than most people? Do fellow walkers sometimes ask you to slow down?

- Driving: Do you get excessively annoyed in slow traffic? When you are caught behind a slow driver, do you sometimes honk or make rude gestures to try to speed them up?

- Schedules: Are you addicted to setting and/or maintaining schedules? Do you allot a specific amount of time for each activity? Do you have a fetish about punctuality?

- Listmaking: Are you a compulsive listmaker? When preparing for a trip, for example, do you make a list of things to do or things to bring?

- Nervous energy: Do you have excessive nervous energy? Are you a person who becomes irritable when you sit for an hour without doing something?

- Waiting: Do you get more annoyed than most people if you have to wait in line for more than a couple of minutes at the bank, a store, or to be seated in a restaurant? Do you sometimes walk out of these places if you encounter even a short wait?

- Alerts: Do others warn you to slow down? How often have you heard your friends or spouse tell you to take it easier, or to become less tense?

Nearly everyone exhibits time urgency on at least some of these questions. But if your answers indicate an overconcern with time and speed in most or all of the categories, or if you are particularly extreme in even a few areas, then you would probably be classified as a time-urgent personality.[31]

When the sense of time urgency becomes extreme and habitual—when people feel compelled to rush even in the absence of real external time pressures—it may lead to what cardiac psychologists Diane Ulmer and Leonard Schwartzburd call "hurry sickness."[32] If you are curious whether your case has progressed to this advanced stage, look for these three symptoms:

Do you notice . . .

- ... deterioration of the personality, marked primarily by loss of interest in aspects of life except for those connected with achievement of goals and by a preoccupation with numbers, with a growing tendency to evaluate life in terms of quantity rather than quality?

- ... racing-mind syndrome, characterized by rapid, shifting thoughts that gradually erode the ability to focus and concentrate and create disruption of sleep?

- ... loss of ability to accumulate pleasant memories, mainly due to either a preoccupation with future events or rumination about past events, with little attention to the present? Focusing on the present is often limited to crises or problems; therefore memories accumulated tend to be of unpleasant situations.

Ulmer and Schwartzburd have found that "yes" answers to these questions warrant a diagnosis of hurry sickness. People with this "disease" suffer a wide range of difficulties, ranging from health problems, particularly those related to the cardiovascular system, to the fragmentation of social relationships and to a low sense of self-worth.

But the concept of hurry sickness vastly overgeneralizes the consequences of living life at a fast tempo. There is a saying that to a hammer, everything looks like a nail. And if you are a cardiac psychologist, you see behavior through the template of disease. A rapid tempo in itself, however, does not necessarily spell disease. The relationship between time pressure, time urgency, and hurry sickness is no more single-arrowed for individuals than it is for cultures: external time pressure does not always lead to a sense of time urgency, nor do either of the two invariably produce the symptoms of hurry sickness.

Tempo cannot be reduced to simply the presence or absence of a problem. My students and I have developed a test that measures individual differences in tempo in a broader sense.[33] We have found that the tempo of personal experience separates into five different categories, only one of which is time urgency. When asked about the tempo of their lives, people consider questions about time urgency but they also focus on the speed they perceive in their workplace, the speed they perceive outside their workplace, the level of activity they prefer in their lives, and the tempo they prefer in their surrounding environment.

It is telling that people's responses on any one of these temporal matters are not very predictive of their responses on the others. This suggests that each of the five categories are distinct facets of the general tempo of life people experience. Most important, we have found that fast or slow tempos per se may have less to do with developing hurry sickness than the fit between the personal temperament categories and those concerning physical realities. For example, people with high activity level preferences tend to actually be better off with speedy lifestyles and environments. Also, the balance between the tempo people experience inside the workplace compared to that in their personal lives may be more important for their psychological and physical health than whether they work in a highly time-pressured job or a more relaxed one.

To assess the tempo of your life within this larger picture, you might ask yourself these additional questions:

Do you feel that the tempo of life is too fast, too slow, or just right when it comes to . . .

- . . . your school or work life?

- . . . the city or town where you live?

- . . . your home life?

- . . . your social life?

- . . . your life as a whole?

You don't need a psychologist to interpret your answers to these questions. The fact is that what is too fast for one person spells boredom for another. And the pressure of one moment can be a stimulant at the next. For every Charles Darwin ("A man who wastes one hour of time has not discovered the meaning of life") we hear from an Oscar Levant ("So little time, so little to do"). If left alone, would you take your life at a leisurely tempo? Do you find today's rapid tempo stimulating? Do you feel like people are always rushing you along and making you do things faster than you would like to? Does your work or school often demand that you put in more time than you prefer? Do you like the energy and excitement of big cities, or, if you had your way, would you prefer to live in a slow-paced environment?

There is no inherent good or bad to an individual tempo. What we make of time is a very personal matter.

BEYOND TEMPO

What then, is time? If no one asks me, I know. If I
wish to explain it to someone who asks, I know it not.
ST. AUGUSTINE, *Confessions*, Book II, Sec. 14.

The terms "tempo" and "pace of life" are sometimes used interchangeably. In fact, the speed of our lives does often color the entirety of our temporal experience. A study by the psychologist Marilyn Dapkus underscores the salience of one's tempo.[34] Dapkus, who was interested in learning what type of concepts people use to describe their experience of time, interviewed a group of adults about the full range of their temporal awareness. She found that people tended to frame their responses in terms of tempo, no matter what area of temporal experience they were addressing. When asked about "change and continuity" in life, for example, a typical response was:

As you get older, people say that time appears to go faster. When my boy lives one extra year, that's 10 percent of his life, but when I live one extra year that's only 2 percent of my life.

When addressing the temporal concept that time is limited, a subject responded:

My husband doesn't feel as rushed as I do; he's more re-laxed, he takes it in stride if time runs out. He can say, "That's that," but I'd be trying to cram in one more thing.

In music, attributes like tempo and rhythm are distinct entities. They can be analyzed independently. In the world of social time, however, the lines are less clear cut.

But the pace of life people experience goes beyond tempo. The pace of life is a tangled arrangement of cadences, of perpetually changing rhythms and sequences, stresses and calms, cycles and spikes. It may be regular or irregular, and in or out of synch with its surroundings. The pace of life transcends simple measures of fast or slow. It is this overlaying and interconnectedness of tempo with the many dimensions of social time, I believe, that constitutes the pace of life that people experience. The chapters that follow explore some of these other facets of the pace of life, beginning with perhaps the closest relative of tempo, the psychological experience of duration.

TWO

DURATION

The Psychological Clock

When you sit with a nice girl for two hours, it seems
like two minutes; when you sit on a hot stove for two
minutes, it seems like two hours. That's relativity.
ALBERT EINSTEIN

I n a 1936 study of time perception, researchers Robert
Macleod and Merrill Ruff confined two subjects to isolation
units in psychological laboratories at Cornell University for
48-hour periods. By the morning of his first day, subject number
one had already become bewildered by his task. At what was actually 7:20 A.M., he wrote the following in his diary:

Oh, the devil! I am losing track of the time again. I must reit-
erate that it really does not interest me. I'll try to keep you
informed as best I can, however, since you want to know. But
please realize that unless I specify otherwise I am guessing
wildly . . . According to my guessing schedule it ought to be
nearly 11:30 A.M.

The second subject was Macleod himself. By that first afternoon, he, too, had lost any objective sense of duration. An excerpt from his journal:

> The last few judgments I have made have been almost perfectly at random. I find that I have lost almost all interest in the problem of the estimation of time. When the signal comes I just make a wild guess.[1]

Duration refers to the time that events last. If we think of tempo as the speed of events, then duration is the speed of the clock itself. For the physicist, the duration of a "second" is precise and unambiguous: it is equal to 1,192,631,700 cycles of the frequency associated with the transition between two energy levels of the isotope cesium 133. In the realm of psychological experience, however, quantifying units of time is a considerably clumsier operation. As Macleod and Ruff learned, when people are removed from the cues of "real" time—be it the sun, bodily fatigue, or timepieces themselves—it doesn't take long before their time sense breaks down. And it is this usually imprecise psychological clock, as opposed to the time on one's watch, that creates the perception of duration that people experience.

Theoretically, a person who mentally stretches the duration of time should experience a slower tempo. Imagine, for example, that baseballs are pitched to two different batters. The balls are thrown every 5 seconds for 50 seconds, so a total of 10 balls are thrown. We now ask both batters how much time has elapsed. Let's say that batter number one (who loves hitting) feels the duration to be 40 seconds. Batter number two (bored by baseball) believes it to be 60 seconds—a range of distortion that is quite common between two individuals, as we will soon see. Psychologically, then, the first person has experienced baseballs approaching every four seconds (10 baseballs / 40 seconds = 4 balls per second) while the second sees it as every six seconds (10 balls / 60 seconds = 6 balls per second). The perceived tempo, in other words, is 150 percent faster for batter number one. As the external clock slows down, so does the experienced tempo.

There is some evidence, for example, that cooler body temperatures can cause people's internal clocks to click at a slower rate. One experiment found that divers immersed in 39-degree sea water estimated a 60-second interval to pass more than 10 percent faster than they did before entering the water. Other studies have found that people with high fevers perceive the clock to move slower than it actually does (they overestimate intervals of time).[2] These findings raise the possibility that people in warmer places are operating on slower internal clocks. This would, in turn, cause the speed of events to seem faster to them, perhaps explaining why their actual temporal norms are kept slower. In other words, in terms of their own internal metronomes, there may be little or no difference in the subjective tempo experienced by people in hotter and colder climates. The tempo in both cases may seem just right.

As we shall see, however, the experience of duration has many permutations. Counting the outward speed of events is, for the most part, a straightforward, objective exercise. But the perception of duration—the denominator of the tempo equation—resides in the realm of subjective experience. The psychological clock, or the speed with which time is perceived to move, is distorted by a host of psychological factors, each of which may have profound effects on how the pace of life is experienced.

THE DISTORTED PSYCHOLOGICAL CLOCK

Man measures time, and time measures man.
OLD ITALIAN PROVERB

Events that last for less than a few milliseconds are perceived as instantaneous, without duration. Beyond these few milliseconds, however, occurrences are subject to conscious experience and memory. They then become framed in temporal units of perceived duration. The experience of duration is multifaceted. We may experience the duration of the moment as it is passing and may then reexperience this same time period retrospectively—what cognitive psychologist Richard Block refers to as "experienced duration" as opposed to "remembered duration." There is considerable evidence that these two visions of time passing not only diverge from one another, but that both are subject to great distortion.[3] They also vary wildly from situation to situation in their degree of inaccuracy, and each individual and culture experiences them very differently.

One of the earliest studies of short-interval time estimations, by the German scientist E. von Skramlik, concluded that the human physiological clock is about 400 times less accurate than the best (by 1930's standards, no less) mechanical timepiece.[4] Although the precision of von Skramlik's estimate may be open to question, there is little doubt that people do, in fact, find it difficult to accurately judge the duration of time intervals. Studies have consistently shown, for example, that the vast majority of people are grossly inaccurate when asked to estimate relatively long-term time intervals. In two typical experiments, researchers found that only about one-quarter of all people were able to accurately judge the passage of periods ranging from one to 25 hours within +/- 10 percent of the actual interval. Without access to timepieces, in other words, three out of four people experience an average day to vary by more than two and one-half hours in either direction.[5]

People also vary in the accuracy of their judgments from one hour to the next. In one experiment, for example, the standard deviation (a statistical measure of variability) of each rater's estimates from one hour to the next ranged from about 25 percent to 49 percent of the 60-minute intervals. Roughly translated, these figures indicate that the average variation in any given person's estimates of consecutive one-hour periods ranged from about 15 to 29 minutes in either direction. Not only, then, did all subjects fluctuate dramatically in their estimates of the duration of different hours, but the fluctuations of some people were almost twice those of others.[6]

All other things being equal, distortions of duration tend to be in the direction of underestimation: the clock moves faster than people think it does. Under "normal" conditions people on the average estimate the passage of an hour as a little more than 67 minutes.[7] The further removed they are from external cues, however, the more exaggerated are their errors. In one study, individuals who were confined to isolation units for periods ranging from one week to one month underestimated the passage of hours by a little under 50 percent. They judged one-hour periods to take an average of a little over one hour and 28 minutes.[8]

The experience of time was even more distorted in an unusual case study by a dedicated French geologist named Michel Siffre. Assuming the role of both experimenter and subject, Siffre confined himself to live alone for two months in an 8 x 13 foot "isolation unit" (a nylon tent) in a cave, on a glacier, 375 feet below the surface of the earth. Not surprisingly, Siffre found that his judgments of the passage of both short and long time periods were radically distorted. For example, he experienced hour-long intervals as lasting, on the average, more than two hours. And he judged that the experiment was only in its thirty-fourth day when he was brought back to the surface after two months. His temporal confusion was expressed in a diary entry only five days into his experiment:

Although it's only six or seven o'clock in the evening, according to my time graph, I'm beginning to yawn. This is ridiculous! It means I am losing a half day every twenty-four hours! Every time I wake up, I am convinced it's too early, that it's only two or three o'clock in the morning. And when I am hungry, I suspect that it's eleven o'clock. And the lapse of time between these two periods (seems) very short . . . Soon after eating a meal I am sleepy; and then I think it must be four o'clock in the afternoon.

By the final few days of his isolation, Siffre had gone off the temporal deep end:

When, for instance, I telephone the surface and indicate what time I think it is, thinking that only an hour has elapsed between my waking up and eating breakfast, it may well be that four to five hours have elapsed. And here is something hard to explain: the main thing, I believe, is the idea of time that I have at the very moment of telephoning. If I called an hour earlier, I would still have stated the same figure.[9]

It would be difficult to argue with Siffre's observation that he was "losing all notion of time."

The recall of shorter intervals often suffers as much or more distortion in some contexts. For example, estimations of the duration of crimes—a situation where temporal accuracy can be critical—is plagued by imprecision. In one experiment, students on a university campus observed a faked assault lasting 34 seconds. When questioned later, they estimated the crime to have endured, on the average, 81 seconds—an overestimation approaching 250 percent.[10] In another series of experiments, memory researcher Elizabeth Loftus and her colleagues asked people to watch a short videotape of a bank robbery and, 48 hours later, to estimate the duration of the tape. They found that, on the average, observers described the 30-second tape as having lasted for about 150 seconds—an error on the order of 500 percent. Only two of their 66 subjects underestimated the duration of the event. And although both genders were vastly more likely to overestimate than underestimate, women were even less accurate than men—they estimated the interval to take about 50 percent longer than did males.[11]

There are situations where the accuracy of people's estimates of the duration of crimes have serious ramifications. Loftus, for example, describes a 1974 case in which the prosecutor charged first-degree murder while the defense claimed self-defense:

The trial arose out of an incident in which a heated argument began between the woman and her boyfriend; she ran to the bedroom, grabbed a gun and shot him six times. At the trial a dispute arose about the time that had elapsed between the grabbing of the gun and the first shot. The defendant and her sister said 2 seconds, while another witness said 5 minutes. The exact amount of elapsed time made all the difference in the world to the defense, which insisted the killing had occurred suddenly, in fear, and without a moment's hesitation.[12]

How commonly do eyewitnesses' overestimations of duration occur in the real world? In one resourceful study, a group of researchers took advantage of data from a crime victimization survey conducted by the city of Portland, Oregon. Part of this survey had asked respondents to estimate how long it took police to arrive on the scene. The researchers were then able to match respondents' recollections of these time intervals, concerning a total of 212 crimes, with official police records of the same incidents. Official police reports are considered highly accurate, and are recorded by dispatchers to the nearest second. Like Loftus, the researchers found that virtually all subjects (all but two) estimated that the police had taken longer to arrive than they actually had. About half of these estimates were off by more than 15 minutes. Perhaps most remarkably, close to 10 percent of the victims overestimated by two hours or more.[13]

Individual differences add even further muddle to the accuracy of time perception. We know, for example, that extroverts are more accurate time estimators than are introverts[14], obese people are more accurate than normal weight people[15], and that heavy drug users tend to be more accurate than light drug users[16]. We also know that psychological time moves more rapidly than that on the clock for manics, hysterics, psychopaths, delinquents, and paranoid schizophrenics, while it seems slower than the clock to melancholics, neurotic depressives, people with anxiety reactions, and nonparanoid schizophrenics.[17] There is even some empirical evidence for the popular truism that time passes more quickly as we age; older people may be interested to know that even college students make this claim.[18]

Stretching Time

A mind that is fast is sick. A mind that is slow is
sound. A mind that is still is divine.

MEHER BABA

The subjectivity of the psychological timekeeper is not always a
flaw. For some people, the distortion of duration is a cherished
skill. It is an active and conscious strategy for controlling the pace
of events.

To the Buddhist master, for example, it is said that a moment
may be eternal. One of the primary tasks of Zen is learning to experience the here and now so utterly that time appears to stand
still, to be "liberated from time," as Alan Watts has said. Some martial arts masters are known for their ability to psychologically
stretch the instant. "At the moment of life and death," says Zen
scholar D. T. Suzuki, "what counts most is time, and this must be
utilized in the most effective way."[19] Martial artists control the moment by learning to psychologically slow down the speed of an opponent's movements, so that the attack appears to come in slow
motion. By doing so, they are able to carefully attend to each significant detail of the impinging event. The master picks off each
threat, one at a time, as if it were standing in wait. Psychologist
Robert Ornstein calls instances like these, where people are able
to experience action as if it were occurring in the infinite present,
as living "in-time."

Contemporary Western athletes speak in their own Zen-like
terms about time expansion. Tennis great Jimmy Connors has described transcendent occasions when his game rose to a level
where he felt he'd entered a "zone." At these moments, he recalls,
the ball would appear huge as it came over the net and seem sus-
pended in slow motion. In this rarified air, Connors felt he had all
the time in the world to decide how, when, and where to hit the
ball. In truth, of course, his seeming eternity lasted only a fraction
of a second. Basketball chatter is also laced with mystical-sounding

references to "getting into zones" where time stands still. Players describe unexplainable occasions when everyone around them seems to move in slow motion. During these moments they report a feeling of being able to move around, between, and through their opponents at will.

In football, former all-pro quarterback John Brodie recalls how, in the most intense moments of a game, "time seems to slow way down, in an uncanny way, as if everyone were moving in slow motion. It seems as if I had all the time in the world to watch the receivers run their patterns, and yet I know the defensive line is coming at me just as fast as ever."[20]

Former Grand Prix champion racing car driver Jackie Stewart believes that successful performance in his high-speed sport requires moving the elements into slow motion:

> Your mind must take these elements and completely digest them so as to bring the whole vision into slow motion. For instance, as you arrive at the Masta you're doing a hundred and ninety-five mph. The corner can be taken at a hundred and seventy-three mph. At a hundred and ninety-five mph you should still have a very clear vision, almost in slow motion, of going through that corner-so that you have time to brake, time to line the car up, time to recognize the amount of drift, and then you've hit the apex, given it a bit of a tweak, hit the exit and are out at a hundred and seventy-three mph.[21]

The ability to stretch time isn't limited to Buddhist masters and gifted athletes. A number of psychological studies have demonstrated that time expansion is well within the reach of common mortals. Some particularly provocative findings have come from research on hypnosis. Psychologists Philip Zimbardo, Gary Marshall, and Christina Maslach, for example, gave hypnotized college students the simple suggestion to "allow the present to expand and the past and future to become distanced and insignificant." These instructions led to dramatically increased absorption in the present moment in people's language, feelings, thought processes, and sensory awareness on virtually every task that they were asked to undertake.[22]

"Expanded present" subjects don't only feel more immersed in the here and now; there is also evidence that hypnotically induced time expansion may result in greater accomplishment per unit of real clock time, just as the masters of Zen and athletics are able to achieve on their own. In one series of studies, for example, hypnotized individuals were offered suggestions such as "Now I'm going to give you much more time than you need to do this experiment. I will give you twenty seconds world time. But in your special time, that twenty seconds will be just as long as you need to complete your work. It can be a minute, a day, a week, a month, or even years. And you will take all the time you need." One hypnotized subject in this study was a secretary who had been interested in designing dresses but was previously unsuccessful in this pursuit. In the course of two half-hour periods in the waking state she was unable to develop any designs. Given time stretching instructions, however, she was able to produce several skillful designs during intervals lasting less than a minute. Psychologically, she experienced these brief periods of clock time as an hour or more. Similarly, a professional violinist reported that she was able to use her subjectively expanded time to practice and review long musical pieces. She later reported that the extra time improved her memory and her technical performance.[23]

The time stretching potential of hypnosis has been well replicated. Aldous Huxley, who had a great deal of experience with hypnosis (and, perhaps not incidentally, with psychedelic drugs), described in his book *Island* how a person in a deep trance[24]

> . . . can be taught to distort time. One starts by learning how to experience twenty seconds as ten minutes. A minute is a half hour. In deep trance, it's really very easy. You listen to the teacher's suggestions and you sit there quietly for a long, long time. Two full hours. You are ready to take your oath on it. When you've been brought back you look at your watch. Your experience of two hours was telescoped into exactly four minutes of clock time.

But slow-moving time is not always a gift. As everyone knows, too much time can be extremely oppressive. When the duration feels *too* slow, life is experienced as simply boring. As the speed of time descends below a critical point—what personality psychologists refer to as one's "optimal arousal level"—the clock often seems to drag.[25] This boredom may then perpetuate itself as a self-fulfilling prophecy; by its very definition, one of the characteristics of boredom is a lack of interest in whatever is occurring, which, in turn, drains us of the energy required to create the very stimulation needed to push the speed of time passing to a more agreeable level.

What determines whether the slowing of time is an empowering experience or—as we used to say in the 1960's—a drag? In large part, the difference between the martial artist's slow-motion world and that of boredom narrows down to a matter of perceived control. The martial artist controls the speed of events. He or she slows down the external world in order to take charge of what would otherwise be too complex to take on. The resulting sense of competence is exhilarating. In Zen, the extreme slowing of time is a complete sense of timelessness—literally, *nirvana.* In the case of boredom, the slowing down of the clock feels outside of one's control. The boredom controls the individual's sense of time. In the martial artist's zone, everyone else seems to slow down, but one's own clock remains at full speed. In the case of boredom, it is the time of inner experience that becomes depressed. Time moves slowly in both the internal and the external world. Affectively, this slowing down is experienced at the very least as unpleasant; often it is extremely painful.

At the pathological extreme of boredom lies a sense of hopelessness. Clinically depressed individuals often describe their pain with the very same words spoken by the martial artist—that each moment feels like eternity. For the depressed person, however, the stretching of time is a chilling experience. One depressed patient described how "the future looks cold and bleak, and I seem frozen in time."[26] The mental slowing of the depressed patient in itself leads to a downward spiral. The slowing impairs skilled action, which causes hopelessness about the future, which may lead the person to give up trying, all of which leads to further mental slowing. At its worst, the belief that there is no future, that the pain of the present is eternal, may lead to suicide. The psychiatrist Frederick Melges believed that this loosened connection to the future is what makes the time of depression pass so slowly. He argued that the fundamental task for treating the hopelessness of depression is to "unfreeze the future." Temporal anxiety is also often common in schizophrenics. Melges reported how the pain of one acute schizophrenic patient was expressed in the statement "Time has stopped; there is no time. . . . The past and future have collapsed into the present, and I can't tell them apart."[27] The slowness of time passing—its duration—has become unendurable.

Herbert Spencer once defined time as: "That which man is always trying to kill, but which ends in killing him." It seems one of life's more curious ironies that time—the most precious of all resources, and surely the least replaceable—is not always welcomed as a gift.

FIVE INFLUENCES ON THE PSYCHOLOGICAL CLOCK

> VLADIMIR: That passed the time.
> ESTRAGON: It would have passed in any case.
> VLADIMIR: Yes, but not so rapidly.
> SAMUEL BECKETT, *Waiting for Godot*

There are at least five major factors that influence the experience of duration. People tend to perceive time passing more quickly when experiences are pleasant, carry little sense of urgency, when they are busy, when they experience variety, and during activities that engage right-hemisphere modes of thinking. But there are vast cultural and individual differences in how each of these influences are interpreted. The same activity may seem like a moment to one person but an eternity to the next.

Pleasantness

Experiments bear out the wisdom of the old saying, "Time flies when you're having fun." One of life's crueler ironies is that time seems to drag when we wish it to go by quickly and it goes by quickly when we want it to be savored. Studies have shown, for example, that people estimate experiences where they succeeded as significantly shorter than those where they failed.[28]

A number of cognitive psychologists have tried to explain this phenomenon. Robert Ornstein believes that the perception of duration is determined by how much we experience and remember of a situation. A successful experience, he argues, is better organized in memory than a failure. Better organized memory packages result in smaller storage sizes, which are perceived as shorter in duration.[29] In other words, our memories of good experiences take up less cortical space and, consequently, are experienced as having taken less time.

Richard Block and his colleagues have proposed a refinement to Ornstein's model that they refer to as the "contextual-change" model. They have demonstrated that it may not be storage size per se that is the critical factor in remembered duration, but the number of changes that occur in the cognitive context—for example, what people are doing or what is occurring in the background.

The reverse also appears to be true. When time is made to pass quickly, people usually perceive what they are doing to be more pleasant. Psychologists and planners have sometimes used the "time flies" phenomenon to their advantage. In one project, for example, psychologist Robert Meade was able to improve workers' morale by speeding up the psychological clock. Meade took advantage of the fact that time is experienced as shorter when people believe that they are making progress toward a goal. This sense of progress, he found, can be enhanced through simple procedures such as establishing a definite end point to the task and providing incentives to reach these goals. Before his experiment, Meade heard comments from workers like "It seems like the day would never end," or "It seems like I've been here all day but it's not even lunchtime yet." After establishing a sense of progress there were proclamations such as, "The day went by so quickly—it seems like I just got started." It is difficult to know, of course, to what extent speeding up the passage of time led to a more pleasant experience or vice versa. The direction of cause and effect, however, is less important than the net effect on workers' well-being. Employers might be pleased to note that these increases in morale are often also accompanied by accelerated production.[30]

Degree of Urgency

The greater the urgency, the more slowly time passes. To a parent with an injured child, the trip to the hospital seems endless. The infatuated lover compulsively counts the passing minutes while awaiting his paramour's return. The woman who yearns to have a child may become obsessed by her biological clock as she sees midlife approaching. The urgency rule extends to a wide range of needs: everything from basic physiological demands to those generated by the broader culture.

Salespeople are well aware of the power of creating a sense of urgency. Limited-time offers are a mainstay marketing strategy. Stores regularly offer one-day or one-hour or even five-minute markdowns. Automobile salespeople often present customers with offers that expire when they leave the premises. Movie advertisers are particularly fond of creating a sense of urgency. One recent movie theater sign invoked time urgency three separate times in a five-word announcement: "Exclusive, limited engagement ends soon!"[31]

My all-time favorite billboard was an advertisement for Pepto-Bismol. The huge sign used to sit above the Santa Ana Freeway about 15 miles outside of downtown Los Angeles. Pictured in the center of the billboard was a gruesomely nauseated, pained and anxious-looking Gahan Wilson–type figure hunched over the steering wheel of his car, driving in the same Santa Ana freeway traffic. At the bottom of the cartoon in large jagged letters were the words: "Diarrhea? The last 15 minutes are the toughest." Under that was the pitch for a smiling bottle of Pepto-Bismol: Never leave home without it.

Cultures have their own norms for handling urgency. As Americans traveling abroad are acutely aware, their own attitudes often conflict with those in most other countries, even those of their first-world colleagues in Western Europe. In most matters, Americans tend to require greater urgency before feeling compelled to discharge their tension. To their credit, people in the United States typically score high on measures of "delay of gratification," a trait that psychologists usually equate with emotional maturity and achievement. On the down side, however, this unwillingness to deal with immediate needs can result in unnecessary unpleasantness. At the risk of fixating on matters of the toilet, an observation from anthropologist Edward Hall seems worth noting:

> The distribution of public toilets in America reflects our tendency to deny the existence of urgency even with normal physiological needs. I know of no other place in the world where anyone leaving home or office is put to periodic torture because great pains have been taken to hide the location of rest rooms. Yet Americans are the people who judge the advancement of others by their plumbing. You can almost hear the architect and owner discussing a new store's rest room. Owner: "Say, this is nice! But why did you hide it? You'd need a map to find it." Architect: "I'm glad you like it. We went all out on this washroom, had a lot of trouble getting that tile to match. Did you notice the anti-splash aerated faucets on the wash basins? Yes, it would be a little hard to find, but we figure people wouldn't use it unless they had to, and then they could ask a clerk or something."[32]

What one means by "Now!," in other words, may require more than a dictionary to translate.

The Amount of Activity

Instead of saying "Don't just sit there; do something," we should say the opposite, "Don't just do something; sit there."

ZEN MASTER THICH NHAT HANH

As clock- and oven-watchers have long suspected, time seems to move more quickly when a task is absorbing, when it is challenging and requires mental effort, and when more events are happening.[33] One experiment, in fact, literally demonstrated that the passage of time feels slower to people who are directed to do nothing other than wait for a pot to boil.[34] In other words, when the tempo of events is fast, the duration of time also seems to compress. In fact, no single temporal dimension tells us more about the psyche of a culture than knowing what its denizens think about activity and inactivity, and how these events—and non-events—affect the psychological passage of time.

In most of the United States, keeping busy is generally considered a good thing, while doing nothing signals waste and void. Inactivity is dead time. Even leisure time in the United States is planned and eventful. We live in a culture where it is not uncommon for people to literally run in order to relax, or to pay money for the privilege of pacing on a treadmill. It sometimes seems as if life is constructed with the primary goal of avoiding the awkwardness and sometimes the terror of having nothing to do.

In many cultures, however, there is less distinction between being active and doing nothing. In Brunei, people wake up in the morning asking, "What isn't going to happen today?"[35] In Nepal and India, I have watched friends drop by one anothers' homes, only to sit and remain silent—visits during which everyone was perfectly comfortable (other than myself, of course). Sometimes the silence would extend for hours, at which time a conversation, often full of laughter, would explode as if through spontaneous combustion. Then there would again be silence, which might continue until it was time to leave. These people were confused when

I asked them whether they felt awkward about doing nothing together. Simply sitting, they explained, was doing something.

Writer Eva Hoffman described how, during a recent long excursion through Eastern Europe, she came to appreciate people's acceptance of silences:

> Now again we wait, facing each other silently. Balkan time. We sit, the way Zen masters sit. There's no awkwardness in it, no frantic noddings of the head or reassuring smiles. I'm beginning to find it strangely relaxing. I'm shifting to another sense of events, in which you don't insist on fulfilling a plan, but wait for what happens next.[36]

Over the course of her journey, Hoffman began to understand that trusting silence requires a faith in the dynamics of change, and in human nature itself:

> Something always happens next: the principle I've been slowly soaking in. The world doesn't run out, and neither do human beings, who for the most part are a source of help rather than threat.[37]

When silence is valued, it ceases to be wasted time. It no longer drags on the clock.

In some cultures, doing nothing is highly treasured. It is not seen as merely a break in the action, but as a productive and creative force. The Japanese, for example, hold the concept of *ma*—roughly meaning the spaces or intervals between objects and activity—in the highest esteem. Westerners might refer to the space between a table and a chair as empty. The Japanese refer to the space as "full of nothing." What doesn't happen for the Japanese is often more important than what does occur, a concept that often baffles Western visitors. For example, comprehending the meaning of verbal communication in Japan requires attending to what is not said, sometimes more than to what actually is said. Because of this, simply understanding the difference between "yes" and "no" can make a naive *gaijin* (foreigner) have a fit. Although the Japanese have a definite word for "no" (*iie*), it is rarely used. Most questions, it turns out, are answered "yes" (*hai*), or not answered at all, no matter whether the respondent means "yes" or "no." As Keiko Ueda points out in her essay "Sixteen Ways to Avoid Saying 'no' in Japan,"[38] the Japanese are taught that it is usually rude to confront questioners with a direct denial. Instead, the questioner is expected to listen to what is not said. Usually, this nonverbalized "no" is expressed in either of two ways. Most common is pausing before responding "yes." The longer the duration of the respondent's pause before offering up a "*hai*," the greater the likelihood that they mean "*iie*." The second, more transparent, approach is by giving no direct answer to the request whatsoever. In both cases it is, quite literally, the silence that carries meaning, while the spoken words are understood to signify nothing.

Marsiela Gomez, a doctoral student in pharmacology at Johns Hopkins, is a part Mayan who was raised in a culture that taught the value of waiting for others to speak first. This habit has often caused problems for her in the United States: "It is very frustrating, because people think I have nothing to add. Sometimes I find that when you wait to speak the answers are upcoming. In this society, it's so important for individuals to own a point of view that everyone feels the need to be the first to put a certain opinion forward. Oftentimes, if I wait long enough, someone will express my point of view." She adds, "Sometimes if one waits too long, the subject changes and then my response is no longer relevant. The need to be heard first seems to be more important than the appropriate response."[39]

Noriko Kito, a psychiatric nurse from Japan who is currently a doctoral student at the Medical College of Georgia, completely understands Marsiela's predicament: "In my country, we don't have to rush to speak. We always have time to think before speaking . . . we have a moment of silence which helps us to process the information . . . And somebody always pays attention to the group, so nobody will be left out from the conversation. It will not happen here. My American friends say that I should be more assertive."[40] A

Japanese friend once told me more bluntly: For Westerners the opposite of talking isn't listening. It's waiting.

To most Westerners, a lack of overt activity signals that nothing is happening. Many people in the world, however, recognize that just because life is quiet on the surface does not mean that change is absent. Periods of nonactivity are understood to be necessary precursors to any meaningful action. The Chinese, for example, are said to be masters of waiting for the right moment. They believe that the wait itself is what creates that moment. How long is the wait? As long as it needs to be. Artificially abbreviating this incubation stage would be as senseless as skimping on the foundation of a building. Helmut Callis has written that "a half century to wait is not too long according to Chinese concepts of time,"[41] an estimate that many Asian scholars consider conservative. Clearly, the absence of surface activity does not have the same meaning in all cultures, nor does it always cause time to move slowly.

Variety

The greater the variety, the more rapidly time seems to move. Lack of variety is a primary component of boredom, which is, by its very definition, the psychological slowing down of the clock.

As is the case for activity, however, Anglo-American standards of what constitutes variety are far from universally shared. Anglo-American culture is addicted to rapid and perpetual change, in everything from fashion to entertainment to the homes and cities where people choose to live. Most people in the world, however, know exactly where they will be living, what job they will be doing, and even what foods they will be eating for the rest of their lives, providing they have any resources at all.

In the United States, today's latest hit, by its nature, becomes tomorrow's throwaway. Fred Turk, who was a colleague during my year in Brazil, is a U.S. citizen who has spent most of his adult life teaching in countries throughout South America. "I don't know if I could ever return to the States," he told me. "I'm always amazed by how alien I feel when I visit. It seems like every time I return people have totally cleared their shelves of yesterday's fashions—not only in clothes, but in music and art and everything else. Even the language seems to change. I never know how to dress, what to talk about or even what words sound foolish. Sometimes, particularly with young people, I can't even follow the conversation."

Turk is describing the U.S. addiction to change that occurs over the time span of weeks and months. An even more dramatic craving for variety may be observed in moment-to-moment shifts. We see this, for example, in the shrinking attention span of television viewers. The popularization of remote control devices and multiple cable stations have produced a generation of what media analysts call "grazers." Recent studies indicate that these viewers change stations as much as 22 times per minute, or once every 2.73 seconds.[42] They approach the airwaves as a vast smorgasbord, all of which must be sampled, no matter how meager the helpings. Compare these grazers to the traditional people of Indonesia, whose main entertainment consists of watching the same few plays and dances, month after month, year after year. Each viewer knows every nuance of movement and each word of dialogue but are perfectly satisfied to return again and again.

Or (taking "smorgasbord" literally) consider the many Nepalese Sherpas whose year-round diet, for their entire lives, consists of the same three meals of potatoes, tea, and after-dinner shots of a potato-based alcohol drink. Once, while traveling in Nepalese villages, I shared a daily diet of potato pancakes for breakfast, boiled potatoes for lunch, and "Sherpa stew" (made from guess what?) for dinner for seven consecutive days. Nobody but me seemed to mind.

Time-Free Tasks

The type of tasks we are engaged in and the nature of the skills that they require may also drastically affect perceptions of duration. The Nobel Prize–winning research of biopsychologist Roger Sperry and his colleagues at the California Institute of Technology have demonstrated that the two hemispheres of our brain tend to focus on different types of information and to process this information in different ways. The left-hemisphere way of knowing is characterized by verbal, analytical thought. It is best at tasks that require labeling, counting, planning step-by-step procedures, making rational statements based on logic, and marking time. The right-hemisphere way of knowing is nonverbal. It is intuitive, subjective, relational, holistic, and time-free. In a loose sense, we have two types of consciousness, what might be labeled left- and right-hemisphere consciousness. As Jerre Levy, one of the foremost scholars in the area of brain asymmetry, states, "The left hemisphere analyzes over time, whereas the right hemisphere synthesizes over space."[43]

The notion of time-free thinking refers to occasions when people completely lose track of time. When engaging in activities that focus on right hemisphere–type thinking, people have difficulty judging duration. It is not so much that the time on the clock mentally speeds up (although it does) but that this mental state seems to exist outside of time. No matter how fast the pace of life unfolds around us, entering the time-free mode puts us at rest, temporally speaking.

For most people, this time-free thinking tends to occur in nonverbal activities such as artwork or music: tasks that require attending to the arrangement of elements in space and seeing how those parts go together to make up the whole. It is particularly likely to appear during tasks that force us to leave verbal, analytical thought behind. Many art instructors, for example, believe that self-convinced drawing incompetents ("But I can't even make a straight line") fail because they approach drawing in the L-mode, when artistic competence requires R-mode thinking ("R-mode" and "L-mode" refer to the perceptual styles that typify the two brain hemispheres). In order to teach these artistic hopefuls to *see* through the proper mode—and nearly all artists agree that creating visual art is primarily an exercise in seeing—popular teachers like Betty Edwards have developed exercises that force students out of the logical, verbal, analytical mode and, consequently, into the R-mode of seeing.[44] Examples of these exercises are upside-down drawing (copying a photo seen upside down) and being asked to draw negative space (the space between objects rather than the object itself). These tasks resist the analytical stereotyping of L-mode thinking and require the student to see every element as it truly exists, without preconceptions. Edwards instructs students that one way to know they are seeing correctly, and are on the road to artistic competence, is when they lose track of time.

Perhaps the epitome of time-free thinking that characterizes the R-mode is what psychologist Mihaly Csikszentmihalyi calls "flow."[45] The experience of flow is the state of consciousness in which one is completely absorbed in the activity at hand. During flow, people exist seemingly outside of time and outside of themselves. Csikszentmihalyi discovered the flow concept by observing artists who spent inordinate periods of time on their work with what appeared to be absolute focused concentration. Madeleine L'Engel compares the artist's focus to child's play: "In real play, which is real concentration, the child is not only outside time, he is outside *himself.* He has thrown himself completely into whatever it is that he is doing . . . his *self*-consciousness is gone; his consciousness is wholly focused outside himself." As the great psychologist Abraham Maslow, who developed the concept of self-actualization, once said about creative people in general, they are "all there, totally immersed, fascinated and absorbed in the present, in the current situation, in the here-now, with the matter-in-hand."[46] They also lose track of passing time. "When consciousness is fully active and ordered," Csikszentmihalyi has found, "hours seem to pass by in minutes, and occasionally a few seconds stretch out into what seems to be an infinity. The clock no longer serves as a good analog of the temporal quality of experience."[47]

It is also clear that some cultural groups tend to engage in the R-mode of thought more than others. The Balinese, for example, have been labeled right-hemisphere people, while Anglo-Americans in the United States lean more toward the characteristics of L-mode thinking. These differences are plainly reflected in attitudes toward time. It is telling that the Balinese—whose daily activities are entwined with religious, musical, dramatic, and artistic ritual—often refer to the hour on the clock as "rubber time" (*jam kerat*). For example, when one asks a Balinese dispatcher, "What time is the bus scheduled to leave?," a typical answer might be "Four o'clock, rubber time."

Life that is filled with a time-free flow mode may take on something of a gamelike appearance. Csikszentmihalyi notes that this occurs "when a culture succeeds in evolving a set of goals and rules so compelling and so well matched to the skill of the population that its members are able to experience flow with unusual frequency and intensity . . . In such a case we can say that the culture as a whole becomes a 'great game.'"[48] A fine description of life in Bali.

There is also, of course, a world of differences, within and between cultures, in the types of tasks that are most likely to trigger this time-free mode. There is some evidence, for example, that music is more clearly associated with the left-hemisphere mode in Japan, while the reverse is true in the United States. It has even been suggested that the type of music makes a difference; that more structured pieces, such as the Baroque concertos of Vivaldi, tend to require more of the left-hemisphere (in-time) mode, while more impressionistic music, such as that of Ravel and Debussy, is more tied to the right-hemisphere (time-free) mode.

BUMPS IN TIME

> Time is a stretch of nerve fibers: seemingly continuous from a distance but disjointed close up, with microscopic gaps between fibers. Nervous action flows through one segment of time, abruptly stops, pauses, leaps through a vacuum, and resumes in the neighboring segment.
>
> ALAN LIGHTMAN, *Einstein's Dreams*

There is an ever-changing texture to time. There are occasions when time seems to flow smoothly and evenly, but other moments when it feels rough and choppy, hard or soft, heavy or light. Although the specific nature of these activities may differ across cultures and individuals, it is clear that everyone experiences the passage of some periods of time differently than others.

Physical scientists have established that the time of the universe—its passage from the Big Bang to the present—has not been smooth and continuous. This is contrary to the Newtonian metaphor, which dominates the thinking of Anglo-American culture, of hands moving around a mechanical clock. Time flows neither universally nor smoothly. Fundamental to Einstein's relativity theory is the understanding that time is not absolute. On the subatomic level, it is clear that particles move both backward and forward in time. Every particle, according to field theory, has its own rhythmic patterns of energy, each seemingly dancing to its own rhythm.[49] Perhaps the most mystifying variation of physical time distortion occurs in the infamous black holes of the universe, where, to an observer from outside the hole, time literally stands still.

The new physics describes the movement of physical time as "bumpy." Bumps in space—in space-time, strictly speaking—result from the dynamics of gravitational attraction. The same might be said for the psychological experience of duration in social time. Unlike the apparently unvarying movement of the time on the clock, the flow of psychological time is bumpier at some times and more evenly fluid at others.[50]

In the 1960's, one of the first computer programs designed to decipher the meaning of sentences was asked to analyze that classic of all statements about temporal movement, "Time flies like an arrow." It came up with the following meanings:

1. Time proceeds as an arrow does (that is, quickly or in a direction).
2. One should measure the speed of flies in the same way as one measures the speed of an arrow.
3. One should measure the speed of flies in the same way as an arrow measures the speed of flies.

4. Go and measure the speed of flies that resemble an arrow.
5. Particular kinds of flies, time-flies, are fond of an arrow.

A subsequent computer logician added his own twist: "Time flies like an arrow; fruit flies like a banana."[51]

The psychological experience of duration is hardly less disorienting. What is the speed of time passing? It depends upon who you ask, and where.

THREE

A BRIEF HISTORY OF CLOCK TIME

... and so it goes. And so it goes. And so it goes. And so it goes goes goes tick tock tick tock tick tock and one day we no longer let time serve us, we serve time and we are slaves passing,—bound into a life predicated on restrictions because the system will not function if we don't keep the schedule tight . . .

The Ticktockman: very much over six feet tall, often silent, a soft purring man when things went timewise. The Ticktockman.

Even in the cubicles of the hierarchy, where fear was generated, seldom suffered, he was called the Ticktockman. But no one called him that to his mask.

You don't call a man a hated name, not when that man, behind his mask, is capable of revoking the minutes, the hours, the days and nights, the years of your life. He was called the Master Timekeeper to his mask. It was safer that way.

HARLAN ELLISON, *Repent Harlequin*

There is no more vivid symbol of a rapid pace of life than the moving hands of a clock. The images of silent film comedian Harold Lloyd dangling for twenty minutes from a clock eight stories above a hurrying city street, or of Salvador

Dali's timepieces melting in a surrealistic desert, provide indelible and enduring imprints of time as the great dictator.

In literature, too, clocks have often taken center stage—most often in the same villainous roles. Possibly the most famous sentence ever written about the pace of life—Thoreau's "If a man does not keep pace with his companions, perhaps it is because he hears a different drummer"—was aimed at a society regulated by the clock. Over the years, many of Thoreau's literary successors have been more direct, and more caustic, about the urgency of slaying draconian drummers. More than a century after *Walden*, Nathanael West in *A Cool Million* spoke for many: "Don't mistake me, Indians. I'm no Rousseautic philosopher. I know that you can't put the clock back. But there is one thing you can do. You can stop that clock. You can smash that clock."

From their earliest introduction, mechanical timepieces have been used not only to mark the beginning and ending of activities, but to dictate their scheduling. They regulate the speed of action and, as critics like Thoreau and West feared, the very pace of society. Clock time has revolutionized the cadence of daily life. It requires an uncompromising regularity in the passage of events. To management, it may seem that the repetitive, rhythmic beat of the clock is what drives production. To social critics, on the other hand, it often seems to underlie a vast temporal monotony. Both sides would agree, however, that more often than not, the regularity of the clock has pushed the pace of events faster than ever before; for many, this pace is well beyond their range of comfort.

For most of industrialized society, living by the clock is a given. As one sociologist observed, "Not to have the correct time in modern society is to risk social incompetence."[1]

People may struggle to escape from clock time, at least to vacation from it, but in the end the ticktockman stands plainly invincible, in control of production and progress. This seemingly inevitable fate is captured in the physicist Alan Lightman's description of a fictional town in Italy:

Then, in a small town in Italy, the first mechanical clock was built. People were spellbound. Later they were horrified. Here was a human invention that quantified the passage of time, that laid ruler and compass to the span of desire, that measured out exactly the moments of a life. It was magical, it was unbearable, it was outside natural law. Yet the clock could not be ignored. It would have to be worshiped.[2]

A look into history, however, reveals that the emergence of the clock-time norm is very recent. For most of human civilization there was no way to insure being punctual even if one wanted to be; and even if a person was on time there was no way they could prove it. Contemporary ideas about promptness and life dominated by the clock would have been incomprehensible to the vast majority of our predecessors. The history of events leading to the now familiar deitylike status of clocks and clock time is a case study of the evolution from the time of nature to that of the clock.

A BRIEF HISTORY OF TIMEPIECES

The first grand discovery was time, the landscape of experience. Only by marking off months, weeks, and years, days and hours, minutes and seconds, would mankind be liberated from the cyclical monotony of nature. The flow of shadows, sand, and water, and time itself, translated into the clock's staccato, became a useful measure of man's movements across the planet. . . . Communities of time would bring the first communities of knowledge, ways to share discovery, a common frontier on the unknown.

DANIEL BOORSTIN, *The Discoverers*

Ancient astronomers were able to mark off years and, to some degree, the months. The measurement of uniform hours, however, is a modern invention. The determination of minutes and seconds is even more recent.[3]

One of humankind's greatest discoveries was the sundial or shadow clock. As early as 5,500 years ago, people detected that an upright post would cast a longer shadow when the sun was lower in the sky. The most primitive of these devices consisted of a simple stick—known as a *gnomon* (from the Greek "to know")—stuck in the ground to harness sunlight and shadows. More elaborate structures, such as those found at Stonehenge, allowed people to measure time in meaningful units. For the first time in history, people could not only mark time but could make appointments—for, say, a hand's width after the shadow appears on the second rock. Eventually, a number of carefully calibrated devices were invented to measure the daylight hours. The ancient Egyptians, for example, developed a sundial consisting of an approximately one-foot-long horizontal bar with a small T-shaped structure at one end. The T cast a shadow along the bar, which could then be calibrated to more exactly measure the passage of time. The bar was set with the T facing east in the morning and was reversed around noon so that the T faced west until sunset. One of these devices from the time of Thutmose III (c. 1500 b.c.) still survives.

But assessing punctuality on the sundial—a device that the Greeks referred to as "hunt-the-shadow"—was an inexact matter at best. What happened when the sun went behind a cloud or went down at night? *Absque sole, absque usu* (without sun, without use) read the inscription on one sundial. In this world of imprecise calibrations, clocks could only measure the brightest hours; and even these times were rough estimates. And the idea of making appointments for specified nighttime "hours," of course, was virtually meaningless.

The next generation of timepieces was aimed at measuring both day and night without having to depend on the weather or sun. The first of these revolutionary devices was the water clock. Within five centuries after the first sundials, inventors began measuring the passage of time by the amount of water that dripped from a pot. Water clocks took many forms, but all came down to measuring the amount of water that passed through a hole. One Egyptian version, for example, consisted of an alabaster vessel with a scale marked inside and a single hole at the bottom. As the water dripped out of the hole, the passage of time could be measured by the drop in the water level from one mark to the next.

Water clocks were sometimes quite elaborate. Daniel Boorstin describes one giant version that ornamented the east gate of the Great Mosque at Damascus:

> At each "hour" of the day or night two weights of brightly shining brass fell from the mouths of two brazen falcons into brazen cups, perforated to allow the balls to return into position. Above the falcons was a row of open doors, one for each "hour" of the day, and above each door was an unlighted lamp. At each hour of the day, when the balls fell, a bell was struck and the doorway of the completed hour was closed. Then, at nightfall, the doors all automatically opened. As the balls fell announcing each "hour" of the night, the lamp of that hour was lit, giving off a red glow, so that finally by dawn all the lamps were illuminated.[4]

It required the full-time attention of eleven men to maintain this device.

The water clock had a long, distinguished career. From the time of ancient Egypt until the invention of the pendulum clock, which appeared about 1700, it was the most accurate device for measuring time when the sun was not shining. For most of recorded history, in fact, sundials served to measure the daylight hours while water clocks marked the hours of night. In ancient Rome, sundials were used to calibrate and set the water clocks.

The Romans put a high value on time. Time meant money. Roman lawyers would often plead with judges for another water clock of time to present their client's case. The phrase *aquam dare*, "to grant water," meant to give time to a lawyer, while the term *aquam perdere*, "to lose water," meant to waste time. When a speaker from the Senate went on too long his colleagues would shout for his water to be taken away.

Even in their simplest derivations, however, water clocks suffered from a number of flaws. For one thing, in cold climates the changing viscosity of the water produced distorted measurements. Another chronic maintenance problem was keeping the hole from clogging, or being worn larger. To prevent wear and clogging in their water clocks the Romans fitted their best models with gems—a precursor to the "jewels" used by later clockmakers.

Drawing upon the logic of the water clock, devices were invented that used anything that would flow, consume, or be consumed. Some of the most popular of these measured time by the burning of oil and candles; and, of course, by the flow of sand through an hourglass. The Chinese developed an incense clock. This wooden device consisted of a series of connected small same-sized boxes. Each box held a different fragrance of incense. By knowing the time it took for a box to burn its supply, and the order in which the scents burned, observers could recognize the time of day by the smell in the air.

The earliest mechanical timepieces appeared in Europe around the fourteenth century. These were weight-driven devices that were neither directed at very small units of time nor capable of measuring them. They were, for the most part, no more accurate than water clocks. The first clocks were invented for one very specific purpose: to inform pious monks when it was time to pray. Before this new invention, monks had mostly relied on hourglasses, which suffered from the inconvenience of needing to be turned regularly. In some monasteries, in fact, a designated monk had to stay up all night to keep the hourglass going until it was time for morning prayers or work. The first clocks were developed to simply sound bells at the appointed prayer hours. Most of these early clocks, which became community centerpieces, didn't even have hands or hours marked on their faces. They were designed not so much to show the time as to sound it. The Middle English word *clok* derived from Middle Dutch and German words for bell. The earliest mechanical timepieces were not technically considered clocks unless they sounded bells. It was several centuries before dials were placed on clocks, and the first dials used only hour hands.

Promptness as defined by an early weight-driven clock—never mind a water clock or an incense burner—would not get very far in today's industrialized world, where dollar values are assigned to hours, minutes, seconds, and even fractions of a second (I recently received a bill for my use of 1.6832 seconds of time on a local computer). When the only measures of time did not have minute hands, promptness as we now know it was clearly not an option. Not until the development of clocks that could accurately measure smaller units of time would the idea of being "on time" or apologizing for being "five minutes late" become meaningful. Before that, telling a friend to "meet me at 5:45" would have been like inviting a person without a calendar to drop by on October 27.

The great breakthrough in timekeeping hardware came toward the end of the sixteenth century with Galileo's discovery of the workings of the pendulum. Galileo learned there is virtual independence between the amplitude of a pendulum's swing and its period of oscillation. A few decades later—around 1700—a Dutch mathematician named Christiaan Huygens developed the first pendulum clock. The best of these early clocks deviated by less than ten seconds per day. Mankind had made remarkable progress measuring the seasons, the weeks, and even the hours of night and day as early as thousands of years ago. But it was only at this point, in the last three centuries, that the pendulum clock offered the potential to live by the precise hour, let alone the minute and second.

It was just after the first mechanical clocks began marking the hours that the word "speed" (originally spelled "spede") first appeared in the English language. Not until the late seventeenth century did the word "punctual," which formerly described a person who was a stickler for details of conduct, come to describe someone who arrived exactly at the appointed time. Only a century after that did the word "punctuality" first appear in the English language as it is used today.

Timepieces have not only evolved toward greater accuracy but have penetrated deeper into our personal space. Helmut Kahlert and his colleagues Richard Muhe and Gisbert Brunner have compiled probably the most comprehensive available discussion of wristwatches in their book *Wristwatches: History of a Century's Development*. They observe how, over history, clocks have kept coming closer to us. Timepieces have progressed from the public clocks of the Middle Ages, to clocks inside the home, to portable pocket watches, to those literally attached to the body. "The wristwatch, apart from the specialized use of the pacemaker, is the last stage of this development, at least for the time being," they say. "It is as near as our skin and always in sight, even at night."[5] Many people take a rather dim view of this development. They might agree with Sigmund von Radecki, who early this century pronounced the wristwatch "the handcuff of our time."

The first armband wristwatches with faces laid out in their now familiar arrangement (the earliest watch faces were set in the position that we would call sideways) began to appear around 1850. For quite some time, though, this design was generally considered to be a failure that would soon pass. Kahlert and his colleagues tell of a professor from Germany who, in 1917, reflected the belief of most everyone in the watch business when he wrote: "The idiotic fashion of carrying one's clock on the most restless part of the body, exposed to the most extreme temperature variations, on a bracelet, will, one hopes, soon disappear."[6] Hardly: as of 1986, it was estimated that the worldwide production of wristwatches was about 300 million per year.

Over the last two centuries the improvements in timepieces have come rapidly. We now live in a world where computers measure time in nanoseconds (a billionth of a second). The National Institute of Standards and Technology in Boulder, Colorado, recently unveiled an atomic clock, NIST-7, that won't gain or lose a second for a million years. The mechanism, they believe, is a significant improvement over its predecessor, NIST-6, which was only guaranteed to remain true to the second for another 300,000 years. (Who, besides physicists, cares about this degree of accuracy? One national newspaper answered this question by pointing to the single fact that "Los Angeles County cuts gridlock by synchronizing traffic lights with NIST–7.")[7]

As physicist Stephen Hawking points out, we can now measure time more precisely than we can measure length. As a result, length is most accurately defined in units of time. The meter may be defined as the distance traveled by light in 0.00000003335640952 seconds. Or, there is a newer and more convenient distance known as a light-second, which is the distance light travels in one second, that may also be used to define both space and time.[8]

For less than the cost of a T-shirt, today's consumer can buy a watch that is calibrated to the hundredth of a second. As a result, public places are now filled with symphonies of little beeps that emanate from watches at the start of each hour. Ironically, our advances in mass precision often magnify the inaccuracies. I've always found it interesting that, although these signals are calibrated to go off with the precision of a hundredth of a second, the beeps almost always sound at different times. When the hour arrives in my lectures, I usually hear watches check in from around the room for a good couple of minutes. Then there's always one deviant who nearly stops me in midsentence with a beep several minutes later.

But for people whose lives are invested in punctuality, today's precision can be remarkable. I once related my observation about unsynchronized hourly beeps to a radio interviewer. He recounted a glaring exception that he had just observed, at a meeting of the National Association of Broadcasters. Virtually everyone at the convention was wearing a watch and nearly every one of these watches had sounded at precisely the same instant at the beginning of each hour. The warble that filled the room, the interviewer observed, was "rather eerie." In a profession where seconds may translate into hundreds of thousands of dollars, the precision of timekeeping is approaching perfection.

THE EARLY INDIFFERENCE TO CLOCK TIME

> Nowadays, in the age of timetables and schedules, it sounds almost funny to learn when reading Heroditus that this great traveler and well-informed man of his age never met the concept of "hour" in his world and could not even find the right word for it. In his time, and even much later, human activity served much more as a measure of time and not the other way around.
>
> ALEXANDER SZALAI

The development of accurate and affordable timepieces provided a technical mechanism that offered the possibility of calibrated, coordinated living.[9] But why did people choose to actualize this potential? How did the clock come to dictate the flow of life, rather than simply serve as an after-the-fact marker of events? Understanding the clock's rise to omnipotent status goes well beyond a description of technological advances. The shift to clock time was brought about by a complex set of economic, social, and psychological forces, and by very aggressive marketing. And the timing of the change was impeccable.

Before the invention of the first mechanical clocks, the idea of coordinating people's activities was nearly impossible. Any appointments that had to be made usually took place at dawn. It is no coincidence that, historically, so many important events occurred at sunrise—duels, battles, meetings.

The historian Marc Bloch tells the story of one of these appointments.[10] There was a medieval duel at Mons that had been scheduled to begin at the usual hour of "dawn." But only one participant showed up. The prompt arriver waited until what he believed to be the hour of "none" (as noon was then called), which marked the required nine-hour arrival deadline. He then asked that his opponent's cowardice be recorded and quickly departed. The only problem was that the umpires, after considerable argument, couldn't agree on whether it was actually "none" when the man left. Eventually, a court had to be convened. After discussion of the evidence—the position of the sun, consultation with clerics who were acknowledged experts in such matters, and vehement debate—the nonetime claim was upheld. The early arriver was officially proclaimed the victor and the no-show was declared a coward.

This incident was typical of what Bloch describes as "a vast indifference to time" within medieval life. Before industrialization, time reckoning was mostly guided by the demands of the environment. Nature dictated when to plant, when to sow, and when to simply sit and do nothing.

The tradition of counting by nature's clocks goes back to the beginning of recorded history. The ancient Egyptian calendar, for example, was a "nilometer"—a vertical scale measuring the rise and fall of the Nile River. Still today, nonindustrial, agricultural societies depend on nature's clock. The Luval tribe in Zambia divide the year into twelve heterogeneous periods of different lengths that are marked by changes in the surrounding climate and vegetation. The Bahan of Borneo divide the year into eight periods, each of which reflect a particular agricultural activity, from the beginning of the brushwood to the celebration of the new rice year.

The Trobriand Islanders of northern New Guinea start their annual calendar, which marks the beginning of the planting season, with the appearance of a worm-the spawning marine annelid, which makes its yearly showing at the southern extremity of the island chain following the full moon that falls between the middle of October and November (our calendar). The Mursi of southwestern Ethiopia also rely on a calendar to direct their agricultural activities. But the Mursi recognize the annual variability of key agricultural events, particularly the onset of the major rainfall period that causes the rivers to flood. As a result, the Mursi treat the calendar as something to be discussed and argued about. They often make up details as they go along.[11]

For a daily clock, most societies have found the sun to be a perfectly adequate marker. The phases of the moon have often been used to mark the months. Native Americans, when they wanted to more clearly distinguish one month from another, sometimes used colorful names for the moons like "Moon when the big trees freeze."

But all this has changed whenever and wherever industrialization and affordable timepieces have come on the scene together. For the most part, to be sure, early clocks were greeted with great enthusiasm. The invention was seen as a liberator from the unreliable time measures that people had previously depended upon. A new class of time emerged: time "of the clock" or "o'clock." Significantly, it was considered a status symbol to make this shift to the time of the clock. Clocks appeared in paintings and in poetry. One medieval Frenchman paid homage in a song:[12]

> The clock is, when you think about it,
> A very beautiful and remarkable instrument,
> And it's also pleasant and useful,
> Because night and day it tells us the hours
> By the subtlety of its mechanism
> Even when there is no sun.
> Hence all the more reason to prize one's machine,
> Because other instruments can't do this
> However artfully and precisely they may be made.
> Hence do we hold him for valiant and wise
> Who first invented this device
> And with his knowledge undertook and made
> A thing so noble and of such great price.

Despite their infatuation with the new invention, however, most people recognized that the truly significant time markers in their lives, those upon which they depended for survival, continued to rest in the hands of nature. Clocks allowed for more precise meeting times; and they were appreciated as exotic ornaments. But the most important temporal events in most people's lives were still agricultural, and nature continued to provide the most precise functional markers for them. Most would have agreed with the words of the modern troubadour Bob Dylan, that "You don't need a weatherman to know which way the wind blows." Until the nine-teenth century, mechanical clocks were considered a poor imita-tion of the time measured by nature.

THE STANDARDIZATION MOVEMENT

> The clock is not merely a means of keeping track of the hours, but of synchronizing the actions of men. The clock, not the steam engine, is the key machine of the industrial age . . . In its relationship to deter-minable quantities of energy, to standardization, to automatic action, and finally to its own special prod-uct, accurate timing, the clock has been the foremost machine in modern technic; and at each period it has remained in the lead: it marks a perfection to-ward which other machines aspire.
>
> LEWIS MUMFORD

But this is getting a bit ahead of the story.[13] Let us go back to the mid-nineteenth century, when the potential for regimentation that clocks offered was still far from a reality. Although timepieces were quickly gaining in quality and numbers, most of life contin-ued to revolve around the cycle of natural events. One of the most formidable stumbling blocks to the acceptance of clock time was the continued lack of standardization of time from one timepiece to another. Clocks were increasingly reliable and available, but their lack of synchronization rendered any advances in precision to be essentially irrelevant. "Counties, provinces, even neighbor-ing villages used different means for telling time," behavioral sci-ence writer Ralph Keyes points out. "In some settings midnight was considered the base hour; in others it was noon, or else sun-rise, or sunset. Even after the invention of mechanical clocks, trav-elers had to reset their clocks repeatedly as they passed from one location to another."[14] In many ways, estimating the time of day was just as chaotic an exercise as it had been in the days of medieval duels.

Historian Michael O'Malley, in his book *Keeping Watch: A History of American Time*,[15] has described an 1843 election dispute that took place in Pottsville, Pennsylvania. Polls were to officially close at 7:00, but many witnesses claimed they saw citizens entering polling places until 8:20. Or had they? "It is well known," claimed the Pottsville *Miner's Journal*, "that we have no exact or certain standard of time in this borough." It was an accepted fact, the editor observed, that the time on watches and clocks in Pottsville typically deviated by "as much as one hour." An election inspector, using a chronometer regulated in Philadelphia three days before, had declared that the polls did indeed close at 7:00. But Pottsville's time was not Philadelphia's time, and the losing side began a campaign to invalidate the election results. O'Malley relates:

> The hearings that followed revealed the multiple sources of time the town used, and the confusion—and political opportunism—that resulted. Several local witnesses deferred to "Heywood and Snyder's Foundry Bell" as their source of time. The bell's prominence made it a common point of reference, but at least one local resident admitted that he routinely set his watch "fifteen minutes slower than the bell" because "I was under the impression that the bell was too fast." One watchless voter testified, "I went down to Geisse [a local jeweler] . . . and at his clock it was twenty minutes past eight." "Henry Geisse's clock," he added, "is always ten or fifteen minutes slower than the [foundry] bell." A watchmaker concurred: Heywood and Snyder's bell was "one quarter of an hour faster" than the sundial he used to regulate his shop. The bartender at the local hotel reported yet another time, about 9:00 P.M., on their "best regulated clock," while Nathaniel Mills insisted that "by my clock the election closed at a quarter after seven."[16]

Which was the correct time? Without an agreed-upon standard, it was impossible to tell.

Well into the nineteenth century, in fact, the world was still blanketed with uncoordinated calendars and time zones. In the United States alone, according to O'Malley, there were about 70 different time zones as late as the 1860's. The Industrial Revolution changed all of this. The new technologies demanded a previously unimagined regimentation and coordination of activities. The clock moved to center stage. By 1880, the number of time zones had dropped to about 50, and scientists were arguing for implementing the total coordination of temporal standards.

Much of the pressure emerged from two particularly frustrated sources: railroad companies and weather forecasters. For the growing network of rail transportation, the lack of coordinated time standards created a nightmare for establishing sensible and efficient timetables. Often, stations just a few miles apart set their clocks by different standards, so that trains were moving backward and then quickly forward in time, at least according to the clocks at each stop. There were frequently two clocks at railroad stations, one exhibiting railroad time and the other local time. During the 1870's the station in Buffalo, New York, for example, had three separate clocks: one for Buffalo city time and the others for the two lines using the station.

Weather forecasters faced a parallel problem. "It was difficult to know how to interpret a weather forecast," O'Malley reports, "if a station in Wisconsin said: 'It's 12 o'clock here, and it's raining.' Those reading the report needed to know whether it meant 12 o'clock by the sun, by Milwaukee time, or by some other standard. The Weather Bureau and the international geophysical community pushed hard for standardization."[17]

Much of the driving force behind these requests for synchronization had resulted from the demands of industrialization. But not all: there were also a few entrepreneurs who recognized the potential of marketing "time" as a product. Two of these individuals, Samuel Langley and Leonard Waldo, were to play particularly pivotal roles in the standardization movement.

Samuel P. Langley, who was to eventually become Secretary of the Smithsonian Institution, was the first to cash in on the growing demand for time coordination. In 1867, Langley took over the directorship of a struggling observatory in Allegheny, Pennsylvania, and quickly developed its timekeeping capabilities. He then persuaded Western Union to connect the observatory to the city. Soon after, he began to literally sell the time, in the form of observatory time signals, by telegraphic transmission to industries throughout Pittsburgh. In 1871, for example, the Pennsylvania Railroad declared Allegheny Observatory time their official standard and contracted for $1,000 a year to receive Langley's signals. Langley also went out of his way to proselytize for standardization, writing a number of articles about the advantages of having a single standard time instead of a myriad of uncoordinated local times. He referred to local time as a "fiction" and a "relic of antiquity," like local weights and measures or local coins, which "the progress of centralization, and the interchange of commerce and travel" had rendered outmoded. Langley established time as a commodity. It was a new twist on the idea that time is money.

A few years later, Leonard Waldo, director of a similar time service at Harvard and later at Yale, went a step further. Waldo took the moral high ground by arguing that time needed to be under the control of scientists. "The furnishing of correct time," he declared, "is educational in its nature, for it inculcates in the masses a certain precision in doing the daily work of life which conduces, perhaps, to a sounder morality." In a report to railroad commissioners about the needs of factory workers, Waldo asserted that "Any service which will train these persons into habits of accuracy and punctuality, which will affect all employers and all employees with the same strict impartiality, so far as wages for time employed is concerned, will be a great benefit to the State."[18] It was the obligation of authorities, he argued, to establish standard time as the regulating authority for what were currently disorganized workers.

Waldo's motives may not have been altogether altruistic. To begin with, he was being paid a salary as director of the Yale Observatory time service. Then, in 1880, he established the Horological Bureau at Yale's Winchester Observatory, which was designed to promote the idea of standard time by rating and testing watches.

Corporate clients were encouraged to send their timepieces there for examination, for which they paid a substantial fee. Finally, in 1882 he formed the Standard Time Company, a joint stock organization that telegraphed accurate time signals, again for a fee, to homes and offices.

In 1883, largely as a result of the campaigns of Langley and Waldo, the railroads established the four time zones used in the United States today. In 1918 the federal government put the four time zones into law, completing the standardization of time in the United States.

MARKETING THE VIRTUES OF CLOCK TIME

The time services of Langley and Waldo and their competitors offered synchronized systems that linked "master" clocks to "slave" or "controlled" clocks located some distance away. As the new system spread, so did the number of local timepieces. The acquisition of timepieces that could link up with the time of the master clock spread first to larger businesses and then to smaller ones. Soon, timepieces were being mass marketed to the public. As Langley and Waldo were selling "the time," clock companies were advancing their own businesses by actively marketing timepieces.

Taking Waldo's lead, one of the primary marketing strategies of the clock companies was to promote the integrity—the inherent superiority—of clock time itself. Advertising campaigns marketed the moral virtue of punctuality. The Electric Signal Clock Company's 1891 catalog, for example, advertised program clocks which could be set to ring bells at preset intervals—the sort of clock-bound bell ringers now commonplace in schools: "If there is one virtue that should be cultivated more than any other by him who would succeed in life, it is punctuality: if there is one error to be avoided, it is being behind time." The company claimed that its best model— vividly named the Autocrat—"gives military precision, and teaches practicality, promptness and precision wherever adopted. A school, office or factory installing this system," the catalog preached on, "is not at the caprice of a forgetful bell ringer, nor anyone's watch, as the office clock is now the standard time for the plant." Not only would the Autocrat standardize time, the brochure explained, but it also offered supervisors a means for extending their disciplinary reach beyond their vision. It "revolutionizes the stragglers and be-hind-time people," the brochure pointed out, because "there is no appeal from these signals—they are the voice of the principal speaking through the standard clock in his office."[19]

The competing Blodgett Clock Company took a similar sales approach: "Order, promptness and regularity are cardinal principles to impress on the minds of young people," argued their 1896 catalog. "No better illustration of these principles than this clock can be secured in a school." Wrapping up their case, the catalog then reprinted a testimonial letter from the principal of a Massachusetts high school who declared that "no assistant in the school is superior to [the Blodgett clock] in promptness and faithfulness . . . I have no hesitation in recommending them to any school officer who is searching for a valuable (I might almost say indispensable) assistant to the school."[20]

In the 1880's a New York jeweler named Willard Bundy and a Scottish physician and mathematician named Alexander Dey independently developed time recording systems that would allow employees to check in and out of work with precision—in other words, to punch a clock. By 1907 nearly all of the leading time card stamping system manufacturers had been bought up by an organization known as the International Time Recording Company, eventually to become IBM. O'Malley reports that they, too, marketed their product by appealing to the virtues of punctuality. The 1914 catalog of the International Time Recording Company argued that time clocks would "save money, enforce discipline and add to the productive time." Also, "the time recorded induces punctuality by impressing the value of time on each individual." The devices, they claimed, improved the character of the factory. "There is nothing so fatal to the discipline of the plant," went another brochure, "nor so disastrous to its smooth and profitable working as to have a body of men irregular in appearance, who come late and go out at odd times." The new time recorders promised to help "weed out these undesirables."[21]

"Keeping a watch" on people became a popular phrase with a new dual meaning. Watchmakers were the term's greatest promoters. The cover on an 1887 pamphlet for the Waterbury Watch Company, for example, announced: "Workingmen Attention: In these Times it is necessary to Keep a Watch on Everybody. For full instructions (on) How to do it, Read This Book." The last page of the booklet showed a policeman with his hand on a man's shoulder, with the latter explaining, "No need to keep a watch on me, Mr. Cop, for I already have the best watch in the world—THE WATERBURY."[22]

The moral gatekeepers of the new industrial society were equally convinced of the virtues of clock time and were more than willing to add their own voices to its promotion. The latecomer was characterized as a social inferior and, in some cases, a moral incompetent.

The urgency of punctual behavior was served up with particularly heavy-handed proselytizing in schoolbooks. An 1881 fifth-grade lesson in *McGuffey's Readers*, for example, began with the scenario: "A railroad train was rushing along at almost lightning speed. The conductor was late, but he hoped to pass the curve safely . . . in an instant there was a collision: a shriek, a shock, and fifty souls were in eternity, and all because an engineer was behind time." The lesson went on to describe a business that failed because its agent was late in making a payment and how an innocent man had been put to death because a messenger carrying his pardon showed up five minutes late. In a grand finale, the lesson asserted that "Napoleon died a prisoner at St. Helena because one of his Marshals was behind time." ("If only Napoleon's marshals could have owned a set of *McGuffey's Readers*," observed Michael O'Malley). The lesson concludes: "It is continually so in life. The best laid plans, the most important affairs, the fortunes of individuals, honor, happiness, life itself are daily sacrificed because somebody is behind time."[23]

The trait of punctuality came to be associated with achievement and success. Living on clock time became a defining characteristic of a new class of people on the move. Owning a watch came to symbolize entry into this fraternity. Historian John Cawelti points out that in Horatio Alger stories the two most crucial events marking the hero's graduation to the middle class are, first, his acquisition of a good suit and, second, his receipt of a good watch. "The new watch," Cawelti explains, "marks the hero's attainment of a more elevated position, and is a symbol of punctuality and his respect for time."[24] Watches became such valued status symbols that some poorer Americans formed "watch clubs." These were essentially watch lotteries, in which each participant put up a weekly fee that went toward the purchase of one new watch. At the end of the week they would draw straws to see who took home the prize.

Even proper maintenance of one's watch became seen as a sign of character. A young man on the way up came to be known as "a real stemwinder." The watch established identity and announced one's social status.

FREDERICK TAYLOR'S EFFICIENCY ENGINEERING

The infatuation with clock time reached a peak with the arrival of Frederick Taylor and his system of "efficiency engineering." Known as the "father of scientific management," Taylor took the clock on a crusade to attain the new Holy Grail of factory management: absolute efficiency. An interesting invention of the scientific management movement, for example, was its motion-and-time studies, the brainchild of Taylor's early disciple Frank B. Gilbreth. This technique involved filming a worker's every movement, with the dual goals of breaking down a company's tasks into their component parts and establishing standard times for each bodily motion. Optimal times, calibrated to fractions of a second, were established for virtually every task. In the process, as Jeremy Rifkin describes,

> The various movements . . . were assigned standard names, using machine terminology. For example, "contact grasp" re-

ferred to picking an object up with fingertips. "Punch grasp" meant thumbs opposing fingers. "Wrap grasp" meant wrapping one's hand around the object. . . . If the task required picking up a pencil, it would be described in the following manner: transport empty, punch grasp, and transport loaded.[25]

After optimum standards were established, the times of each worker's every movement were clocked. The factory owners would then separate "waste" motions—like talking, yawning, scratching one's head or any other "extra" movements—from the movements that led directly to the production at hand. The precision of these measurements was eventually perfected to the level of ten-thousandths of a minute. Taylor believed that his method of scientific engineering, when applied with complete objectivity, would produce the perfectly streamlined "standard time" for every job.

Taylor's technique was then applied to the factory as a whole. After establishing the minimal standard time of each job, each individual stage in the process was linked together in a stepwise sequence under the regulation of a master clock in the central office. When workers began and completed a specific task, they had their card punched by a secondary "slave" clock. These cards went to a "time clerk" in the central planning office, where their elapsed times were compared against the official standard.

Time-and-motion studies have been applied to nearly every work environment. Even the smallest tasks have been targeted for standardized times. A time chart from the Systems and Procedures Association of America, for example, suggests target times for activities like these: open and close file drawer, no selection = .04 seconds; desk, open center drawer = .026 seconds; close center drawer = .027 seconds; close side drawer = .015 seconds; get up from chair = .033 seconds; sit down in chair = .033 seconds; turn in swivel chair = .009 seconds; move in chair to adjoining desk or file (4 ft. max.) = .050 seconds.[26]

Taylorism raised the value of efficiency and the importance of clock time to new levels. Economist Harry Braverman argues that the work of Taylor and his disciples "may well be the most lasting contribution America has made to Western thought since the Federalist Papers."[27] "The new man and woman," observes Jeremy Rifkin, "were to be objectified, quantified, and redefined in clockwork and mechanistic language . . . Above all, their life and their time would be made to conform to the regimen of the clock, the prerequisites of the schedule, and the dictates of efficiency."[28] In time, the stopwatch itself came to stand for Taylorism. Eventually, it would galvanize the enemies of the Ticktockman.

FIGHTING THE NEW TIME

The gods confound the man who first found out
How to distinguish hours—confound him, too,
Who in this place set up a sundial
To cut and hack my days so wretchedly
Into small pieces! When I was a boy,
My belly was my sundial—one more sure,
Truer, and more exact than any of them.
This dial told when 'twas proper time
To go to dinner, when I ought to eat;
But nowadays, why even when I have,
I can't fall to unless the sun gives leave.
The town's so full of these confounded dials . . .
                    (written some two thousand years ago
                    by the Roman playwright Plautus)

By the early twentieth century, particularly in the United States, clock time had been firmly established as the regulator of public life. But not everyone greeted the new time with open arms. Many people understood the profundity of temporal standardization and feared its consequences. They recognized that it established a new conception of time reckoning and, most critically, that it would mean new priorities for the social order. Some of these criticisms targeted the act of standardization; others focused on the more general tyranny and rigidity of the clock.

The act of standardization was met with particularly vocal opposition. From the inception of the standardization movement in the middle of the nineteenth century, there was an uproar from detractors who were unwilling to sanction the new authority of the clock. For example, the New York *Herald* in 1883 observed that standard time "goes beyond the public pursuits of men and enters into their private lives as part of themselves." The *Washington Post* described standardization as "scarcely second to the reformation of the calendar by Julius Caesar, and later by Pope Gregory XIII."[29]

Between 1883 and 1918, when the new time was being enacted by private industry without having been established by federal laws, there were frequent outcries from localities. "Let us keep our own noon," demanded the prestigious Boston *Evening Transcript* as word of the railroad's plan spread. The Louisville *Courier Journal* referred to standardization as "a monstrous fraud," "a compulsory lie," and "a swindle." Michael O'Malley describes a letter to the editor of that newspaper:

> Can you tell me [the letter asked] "if anyone has the authority and right to change the city time without the consent of the people, and what benefit Louisville can derive from it?" The editors responded that no such authority ruled, and no benefit seemed likely from what was "only a disguised step towards centralization . . . a stab in the dark at our cherished State's rights. After they get all our watches and clocks ticking together," the editors asked in reflexive alarm, "will there not be a further move to merge the zone states into districts or provinces?"[30]

Some of the most vocal objections came from the state of Ohio. The Cincinnati *Commercial Gazette*, whose local time was being put back 22 minutes, wrote: "The Proposition that we should put ourselves out of the way nearly half an hour from the facts so as to harmonize with an imaginary line through Pittsburgh is simply preposterous . . . let the people of Cincinnati stick to the truth as it is written by the sun, moon and stars." The *Commercial Gazette*, calling it "a great stupidity" to accommodate the railroad's needs, continued until 1890 to publish railway timetables under the heading: "This is Cincinnati Time. Twenty-two minutes faster than railroad time."[31]

Above and beyond the opposition to standardization, there was the broader issue of the clock's authority over time in daily life. These attacks have been even more sustained and voracious than those against the standardization. It would not be an exaggeration to say that assaults on clock time have often been directed at the fundamental values of modern life itself.

The objections fall into many categories. Some, for example, have been directed at Taylorism and its mechanical approach to human beings. In testimony during Congressional hearings into Taylorism in 1912 one machinist remarked: "I don't object to their finding out how long [a job] takes, but I do object to their standing over me with a stop watch as if I was a race horse or an automobile."[32]

Another type of protest has focused on the controlling aspects of clock time—the Gulliver argument. These critics often question the sanity of a society which allows its existence to be dictated by an artificial creation. Many of the most eloquent attacks on the destruction of natural time have appeared in literature. Kurt Vonnegut, for example, writes in *Slaughterhouse Five*:

> The time would not pass. Somebody was playing with the clocks, and not only with the electric clocks, but the wind-up kind, too. The second hand on my watch would twitch once, and a year would pass, and then it would twitch again.
>
> There was nothing I could do about it. As an Earthling, I had to believe whatever clocks said-and calendars.

And Peter Beagle says in *The Last Unicorn*:

> When I was alive, I believed—as you do—that time was at least as real and solid as myself, and probably more so. I said "one o'clock" as though I could see it, and "Monday" as

though I could find it on a map . . . Like everyone else, I lived in a house bricked up with seconds and minutes, weekends and New Year's Days, and I never went outside until I died, because there was no other door. Now I know that I could have walked through the walls.

An additional target of anger has been the loss of natural time. Editor Charles Dudley Warner, writing just after the invention of standard time in 1884, objected to the rigidity and invariability of time lived by the clock: "The chopping up of time into rigid periods is an invasion of individual freedom and makes no allowances for differences in temperament and feeling."[33]

Norman Mailer has assailed the inorganic monotony of a time where all of life is planned and scheduled; where events are forced to fit the demands of the clock. In *An American Dream*, published in 1964, he equates inorganic time with hell itself. One of his characters comes upon a "vision of hell" when he visits the penthouse of a high luminary in the American power structure:

> . . . a nineteenth-century clock, eight feet high with a bas-relief of faces: Franklin, Jackson, Lincoln, Cleveland, Washington, Grant, Harrison, and Victoria; 1888 the year; in a ring around the clock was a bed of tulips which looked so like plastic I bent to touch and discovered they were real.

Some critics, such as Jeremy Rifkin, believe that the time of computers has hammered the final nails into the coffin of natural time: "The events in the computer world exist in a time realm that we will never be able to experience. The new 'computime' represents the final abstraction of time and its complete separation from human experience and rhythms of nature."[34]

But even within the most clock-oriented cultures, the time of nature still rears its nostalgic head on special occasions. When enveloped by the forces of nature, clock-time people will usually revert back to more "primitive" time-reckoning procedures. The late writer Alex Haley once described how one of the reasons he loved to cruise the sea on small freighters was what it did to his sense of time: "Once you're at sea for a couple of days, time becomes meaningless," Haley observed. "'What day is this?' becomes a frequent query, and the days tend to become identified by their characteristics of weather and sea, or by some special event, such as 'The day after we saw the giant school of green turtles.'"[35] Haley also believed that he did his best writing when in the grip of natural time.

Natural disasters also often lead to similar time-reckoning "regressions." At one point during the devastating floods in the midwestern United States in the summer of 1993, the *New York Times* asked a Missouri resident about the night he was hit hardest. The resident, the newspaper reported, "remembers everything about the night the river forced him and his wife out of the house where they had lived for 27 years—except for this. 'I can't tell you what day it was . . . All I can tell you is that the river stage was 26 (feet) when we left.'" The title of the article was "They Measure Time by Feet."[36]

TIME WARS

> By imposing his own pace, the enemy became time's master, and time itself became our enemy.
>
> ELIE WIESEL, *All Rivers Run to the Sea*

Jeremy Rifkin has described *Homo sapiens* as the only "time binding" animal. "All of our perceptions of self and world are mediated by the way we imagine, explain, use, and implement time," Rifkin observes.[37] Not surprisingly, people react very strongly to intrusions on their understanding of time. The critics of standardization and clock time phrase their attacks in terms of life and death. Beliefs about time reside in a very personal zone, deep inside our psychological spaces, and even minimal meddling may be met defensively.

The uproar caused by standardization and the new clock time in the United States was just one of many time wars. Conflict over the pace of life has been the center of power struggles on many levels. In the personal dimension, for example, there are few greater sources of friction between couples than disagreements about such questions as how to spend their time, when to begin activities and when to go home, who is too fast, who too slow, and who must wait for whom. Jenny Shaw, a sociologist at the University of Sussex in England, asked more than 700 people to write about their experiences with time and punctuality. One of her major findings was the degree of emotional intensity with which people wrote. One woman, for example, complained:

> My husband was lunatic over time keeping . . . would get absolutely furious if anyone was late . . . be really rude and keep up his bad temper for some time. I don't like having to wait about for people but see no point in making a fuss if they are late. But then women usually behave more sensibly and rationally than men if something has gone wrong.

Another respondent wrote about the discrepant psychological clocks of her husband and herself: "in our early married life this, next to lack of money, caused more rows than anything."[38]

This would not have surprised the late psychologist William Kir-Stimon, who devoted much of his professional career to understanding the significance of personal tempo for couples counseling. Kir-Stimon, who was editor of the journal for psychotherapists entitled *Voices*, found what he termed "temporal territoriality" and "tempostasis" frequently produce balanced and synchronous communication in couples. He argued that, along with obvious cultural and familial factors, there is a genetic foundation for individual differences in tempo. Kir-Stimon focused on these temporal conflicts in therapy sessions. He sometimes, for example, brought out a metronome for each member of the couple to set to their preferred cadence. Kir-Stimon would use this information to help couples recognize their differences and to resolve their temporal conflicts.[39] Temporal mismatches may also produce difficulties between parents and children. Even shortly after birth, mothers sometimes experience anxiety if their infant sucks too slowly or too quickly during breast feeding. Unless the mother learns to accept her child's tempo as simply different from her own, she creates tension with her child.

Temporal power struggles are often even fiercer on the larger level of nations and cultures. Historically, religious authorities have taken control of the calendar as a way of asserting and legitimizing their power. Revolutionaries, in return, have often focused their battles for the hearts and minds of the populace on the reigning regime's temporal system. It is a testament to the deep-rootedness of temporal norms, however, that temporal revolutions are almost always short-lived.

A particularly radical attempt at temporal change occurred as part of the French Revolution. In 1793, the French National Convention established a "revolutionary calendar" to replace the reactionary Gregorian one. Among other things, the new calendar declared that: The year 1792 of the Christian era would be the year one of the new Republican calendar; each new year was to begin on September 22 of the old calendar; each month was to be thirty days, with five extra days added at the end of the year; months were to be divided into three ten-day cycles; days were to be broken down into units of ten, rather than 24 hours.[40] It was further pronounced that time from then on would be measured in units of *decades*—decimal minutes and decimal seconds.

The goal of these massive changes, observes sociologist Eviatar Zerubavel, was nothing less than "to gain social control by imposing a new rhythm of collective life."[41] The new calendar, however, met with widespread resistance both within and outside France. One problem was that the ten-day week meant substituting a tenth day for the seventh day of the Sabbath, which meant that the annual days of rest were reduced from 52 to 36. The new calendar also cut the number of holidays under the old system by more than half. For these and many other reasons, the new measure was abandoned after a very uneasy 13 years.

The leaders of the Russian Revolution attempted a similar temporal coup. In 1929 Joseph Stalin, in an effort to do away with the

Christian year, established a revolutionary calendar to replace the Gregorian arrangement. At first, the new system introduced a five-day week—four days of work followed by one of rest, with each month consisting of six weeks. Later, it brought out a six-day weekly cycle. This revolutionary project was abandoned in 1940 when that country, too, returned to the familiar Gregorian calendar.[42]

Jeremy Rifkin predicts that time wars like these will increasingly dominate the politics of the future. "A battle is brewing over the politics of time," he believes. "Its outcome could determine the future course of politics around the world in the coming century." The traditional split in the political spectrum between left and right wings, Rifkin argues, will be replaced by a "new temporal spectrum with empathetic rhythms on one pole and power rhythms on the other." Those aligning themselves with the power time frame are committed to the values of efficiency and speed that characterize the "time is money" dogma of the modern industrial age. Supporters of the empathetic time frame argue against "the artificial time frames that we have created . . . Their interest is in redirecting the human consciousness toward a more empathetic union with the rhythms of nature." Rifkin predicts that "Politics, long viewed as a spatial science, is now also about to be considered as a temporal art."[43]

THE DEMISE OF THE TICKTOCKMAN

Temporal habits die hard. Just as Americans consciously resisted standard time, and the French and Russians resisted the revolutionary calendars, people will continue to battle unsolicited controls on their preferred, "natural" pace of life. Even small changes are quickly noticed.

For the last century and a half most of the clamor has focused on the replacement of the smooth flow of natural time by the discrete, measured moments of the mechanical clock. In recent years, however, changes in timekeeping have come from many directions. One recent challenger to the time of nature, but also to the movement of the mechanical dial clock, is the digital time-piece. It, too, has encountered resistance. Joseph Meeker, writing for *Minding the Earth Quarterly*, for example, offers these reflections on his digital watch:

> However accurate watches and clocks may be, they fail to tell the truth about time. My conventional watch (called these days an "analog" model) is a symbolic arrangement of numbers representing twelve adjacent hours, with continuously moving hands to indicate time passing. When I look at it I see a twelve-hour span, and I learn which part of it I am moving through. The watch measures time by rearranging its objects in space, which is analogous to what the solar system does. The speed of the hour hand is based upon the speed of the earth's daily rotation, so when I glance at my wrist I am reminded that the earth is in motion.
>
> Digital clocks and watches convey no such context. Impaired instruments that they are, they are unable to comprehend more than one instant at a time, with nothing to hint that there is a process going on that includes what went before and what comes after. A digital timepiece resembles a highly trained specialist who has learned to do only one thing, to do it very well, and to ignore all surroundings and relationships. Digital watches and narrow visions fit together very well, and both are signs of our time.[44]

Could it be that the next turn in the evolution of clock time will lead to nostalgia for the Ticktockman? For a return to the *real* time of hour and minute hands? When I lived in Tallahassee, Florida, some years ago, the local government was attempting to pass a law that would convert local time to the surrounding daylight savings time. In opposition to the measure, an angry fundamentalist religious group launched an advertising campaign encouraging people to retain the old standard time. Their slogan was "Preserve God's Natural Time."

How we define and measure our time does, in fact, border on the religious. And people do not change religions lightly.

FOUR

LIVING ON
EVENT TIME

In a world where time cannot be measured, there are
no clocks, no calendars, no definite appointments.
Events are triggered by other events, not by time. A
house is begun when stone and lumber arrive at the
building site. The stone quarry delivers stone when
the quarryman needs money. . . . Trains leave the station . . . when the cars are filled with passengers. . . .
Long ago, before the Great Clock, time was measured by changes in heavenly bodies: the slow sweep
of stars across the night sky, the arc of the sun and
variation in light, the waxing and waning of the
moon, tides, seasons. Time was measured also by
heartbeats, the rhythms of drowsiness and sleep, the
recurrence of hunger, the menstrual cycles of
women, the duration of loneliness.

ALAN LIGHTMAN, *Einstein's Dreams*

Anyone who has traveled abroad—or waited in a doctor's
office, for that matter—knows that the clock, or even the
calendar, is sometimes no more than an ornament. The
event at hand, on these occasions, often begins and ends with
complete disregard for the technicalities of a timepiece. We in the
industrialized world expect punctuality. But life on clock time is

clearly out of line with virtually all of recorded history. And it is not only from a historical perspective that these temporal customs are so deviant. Still today, the idea of living by the clock remains absolutely foreign to much of the world.

One of the most significant differences in the pace of life is whether people use the hour on the clock to schedule the beginning and ending of activities, or whether the activities are allowed to transpire according to their own spontaneous schedule. These two approaches are known, respectively, as living by clock time and living by event time. The difference between clock and event time is more than a difference in speed, although life certainly does tend to be faster for people on clock time. Let me again turn to a personal example.

A few years after my stay in Brazil I became eligible for a sabbatical leave from my university. I decided to invest my term of "rest and renewal" in a study of international differences in the pace of life. I also chose to use the opportunity to live out a childhood dream—to travel around the world.

Precisely where I would go wasn't altogether clear. The phrase "travel around the world" had a lovely ring to it, but I must admit that I wasn't certain just what it entailed. Never having done very well in geography, I had very little grasp of how the nations of the world are arranged and even less notion of their innards. Not knowing what I'd encounter, it was impossible to plan exactly where to visit or how long I would stay in each country. I decided, instead, to let the trip evolve its own form. Fortunately, the research I had designed allowed me the flexibility to decide where and when to collect data along the way.

I bought a map of the world and marked the locations of the four most exotic sights I could invoke: The Great Wall of China, Mount Everest, the Taj Mahal and the Great Pyramids of Egypt. I drew a line connecting the marks. Although I was uncertain how many of these wonders I'd actually see, they gave my trip a rough outline. I decided to fly to the edge of Western Asia and then make my way by land, moving in a rough westerly direction, around the globe. Searching the map for Asia's outside edge, my finger landed on Indonesia.

I purchased a one-way plane ticket to Jakarta, with stops along the way in Japan, Taiwan, and Hong Kong. Beyond that, I had no tickets. From Indonesia I would travel up the Malaysian peninsula toward Thailand, and then west across Asia toward home. My only rules would be to travel no better than second class and to stay on the ground as much as possible. I gave up my house lease, loaned out my car, put my possessions in storage, and told everyone who needed to know I'd be gone for the semester. (Professors don't think in terms of months. Our unit of time is the semester.) The semester stretched into two semesters (one year, tossing in a summer vacation).

The trip began with a flight from San Francisco to Tokyo. Settling in for the long ride, I tried to focus on what I was beginning. My first thought was that I had no keys in my pocket. Next, that in place of an appointment calendar, I was carrying, for the first time in my life, a journal. Then came the realization that I had no commitments. There was nothing, other than carrying out my very flexible research plans, that needed to be done. I didn't have to be any place at any specific time for six whole months. There were no plans or schedules to interfere with whatever might come along. I could let my opportunities come forth on their own and I would choose those I wished to follow. I was free, free, free!

My joy lasted nearly half a minute. Then the terror: What in the world would I do for a whole semester without a schedule or plans? I looked ahead and saw layers and layers of nothing. How would I fill my time? I have never in my life so yearned for an appointment—with anyone for anything. It really was pitiful. Here I was freer and more mobile than most people in the world could ever dream of being. I was Marlon Brando on his motorcycle—with a passport, a Ph.D., and a steady paycheck. And I responded with an anxiety attack.

When I dozed off a little later on the flight, I dreamed about a passage from William Faulkner's *Light in August*. It is when the character named Christmas, hungry and fleeing from the sheriff, becomes obsessed with time. I later looked up the actual quotation:

. . . I have not eaten since I have not eaten since trying to remember how many days it had been since Friday in Jefferson, in the restaurant where he had eaten his supper, until after a while, in the lying still with waiting until the men should have eaten and gone to the field, the name of the day of the week seemed more important than the food. Because when the men were gone at last and he descended, emerged, into the level, jonquilcolored sun and went to the kitchen door, he didn't ask for food at all . . . he heard his mouth saying "Can you tell me what day this is? I just want to know what day this is."

"What day it is?" Her face was gaunt as his, her body as gaunt and as tireless and as driven. She said: "You get away from here! It's Tuesday! You get away from here! I'll call my man!"

He said, "Thank you," as the door banged.

After finally arriving in Tokyo, I checked into a hotel room an ex-student had reserved for me. This was the only room reservation I had for the next six months (twelve months, actually—but, mercifully, I didn't know that then). After unpacking, I put on the robe and slippers provided by the hotel. The bottom of the robe showed considerably more thigh than its maker had intended and the slippers only fit over three of my toes. But I liked the image and, coupled with a dip in the hot tub and a very large bottle of Sapporo beer, I went to sleep with some iota of hope for my immediate future.

The next morning I awoke to a view of green tiled roofs, banyan trees, and an enormous reclining Buddha. At the sight of my little robe and slippers my anticipation returned. I was ready to let events take their own course. What to do first? I loved my hot tub the night before, so decided to start my day with another long dip. Then I found a tea shop next door. The waiter spoke a little English, the food was good, and there was even a *Herald Tribune* to keep me company. After breakfast I explored my neighborhood reclining Buddha, who turned out to be resting in a large temple surrounded by a lovely park. I took out a book to read, stretched my legs and watched life in Tokyo parade by.

Next? A friend had given me a list of gardens he thought I'd enjoy seeing. Why not? I randomly chose one, and thoroughly enjoyed the visit. That evening I had a nice dinner at a restaurant near my hotel. I ended my day with a hot tub, my robe and slippers, and a Sapporo.

The following morning I shot out of bed with an adrenalin charge. What might this new day have in store? How to begin? A hot bath first, of course. Then, recalling the pleasant morning before, I returned to my tea shop for breakfast. After that I could think of no place on earth I would rather be than sitting beside my local Buddha. That afternoon I tried another garden. In the evening I returned to the same restaurant. And, of course, I took a hot bath and nursed my Sapporo before turning in. Another lovely day.

Day three went something like: hot tub/breakfast at the tea shop/Buddha/gardens/dinner/hot tub/Sapporo. The next day was the same. As was the next. And the next.

Looking back at that first week, I see you could have set a clock by my activities. What time is it, you ask? "He's reading his book in the park, so it must be 10 o'clock." "Now he's leaving the hot tub, so that must mean a little after eight." Without intending it, I'd created the structure I so craved on my plane trip. Ironically, one of the very reasons I chose a career in academia in the first place was because it, more than other professions, allowed me to arrange my own time. But when confronted with no limits, I had bounced to the other extreme. To my surprise as well as humbling disappointment, I had built a tighter schedule than the one I lived at work.

DROWNING IN EVENT TIME

My behavior, I now recognize, was a textbook struggle between the forces of clock time, on the one hand, and event time on the other. Under clock time, the hour on the timepiece governs the beginning and ending of activities. When event time predominates, scheduling is determined by activities. Events begin and end when, by mutual consensus, participants "feel" the time is right. The distinction between clock and event time is profound. The sociologist Robert Lauer conducted in his book *Temporal Man* an intensive review of the literature concerning the meaning of time throughout history. The most fundamental difference, he found, has been between people operating by the clock versus those who measure time by social events.[1]

Many countries extoll event time as a philosophy of life. In Mexico, for example, there is a popular adage to "Give time to time" (*Darle tiempo al tiempo*). Across the globe in Africa, it is said that "Even the time takes its time." Psychologist Kris Eysell, while a Peace Corps volunteer in Liberia, was confronted by a variation on this African expression. She describes how every day, as she made her eight-mile walk from home to work, complete strangers would call out to her along the way: "Take time, Missy."

My experiences in Japan were those of a clock time addict floundering in situations where programming by the clock had lost its effectiveness. I was, I have since come to learn, drowning in good company. The social psychologist James Jones had even more complicated temporal challenges during his stay in Trinidad. Jones, an African-American, is quite familiar with the casualness of what used to be called "colored people's time" (CPT). But he was unprepared for the quagmire of life on event time. Jones first confronted the popular motto "Anytime is Trinidad time" soon after arriving, and said he spent the rest of his stay trying to understand just what it meant:

CPT simply implied that coming late to things was the norm and contrasted with the Anglo-European penchant for punctuality and timeliness. Over the course of my year in Trinidad, though, I came to understand that Trinidadians had personal control over time. They more or less came and went as they wanted or felt. "I didn't feel to go to work today," was a standard way of expressing that choice. Time was reckoned more by behavior than the clock. Things started when people arrived and ended when they left, not when the clock struck 8:00 or 1:00.[2]

To visitors from the world of clocks, life conducted on event time often appears, in James Jones's words, to be "chronometric anarchy."

WHERE ARE THE COWS?: MEASURING TIME IN BURUNDI

When event time people do listen to the clock, it is often nature's clock they hear. Salvatore Niyonzima, one of my former graduate students, describes his home country of Burundi as a classic example of this.

As in most of Central Africa, Niyonzima says, life in Burundi is guided by the seasons. More than 80 percent of the population of Burundi are farmers. As a result, "people still rely on the phases of nature," he explains. "When the dry season begins it is time for harvesting. And when the rainy season comes back—then, of course, it's time to return to the fields and plant and grow things, because this is the cycle."

Appointment times in Burundi are also often regulated by natural cycles. "Appointments are not necessarily in terms of a precise hour of the day. People who grew up in rural areas, and who haven't had very much education, might make an early appointment by saying, 'Okay, I'll see you tomorrow morning when the cows are going out for grazing.'" If they want to meet in the middle of the day, "they set their appointment for the time 'when the cows are going to drink in the stream,' which is where they are led at midday." In order to prevent the youngest cows from drinking too much, Niyonzima explains, farmers typically spend two or three hours with them back in a sheltered place, while their elders are still drinking from the stream. "Then in the afternoon, let's say somewhere around three o'clock, it's time again to get the young cows outside for the evening graze. So if we want to make a late appointment we might say "I'll see you when the young cows go out.' "

Being any more precise—to say, for example, "I'll meet you in the latter part of the time when the cows are out drinking"—would be, Niyonzima says, "just too much. If you arrange to come to my place when the cows are going to drink water, then it means it's around the middle of the day. If it's an hour earlier or an hour later, it doesn't matter. He knows that he made an appointment and that he'll be there." Precision is difficult and mostly irrelevant because it is hard to know exactly at what time people will be leading the cows out in the first place. "I might decide to lead them to the river one hour later because I either got them out of the home later or it didn't look like they really had that much to eat because the place where they were grazing didn't have very much pasture."

People in Burundi use similarly tangible images to mark the nighttime. "We refer to a very dark night as a 'Who are you?' night," Niyonzima explains. "This means that it was so dark that you couldn't recognize anybody without hearing their voice. You know that somebody is there but can't see them because it is so dark, so you say 'Who are you?' as a greeting. They speak and I hear their voice and now I recognize who they are. 'Who are you?'–time is one way to describe when it gets dark. We might refer back to an occasion as having occurred on a 'Who are you?' night."

Specifying precise nighttime appointments, Niyonzima says, "gets difficult. 'Who are you?' simply refers to the physical condition of darkness. I certainly wouldn't give a time-like 8 P.M. or 9 P.M. When people want to name a particular time of the night, they might use references to aspects of sleep. They may, for example, say something occurred at a time 'When nobody was awake' or, if they wanted to be a little more specific, at the time 'When people were beginning the first period of their sleep.' Later in the night might be called 'Almost the morning light' or the time 'When the rooster sings'; or, to get really specific, 'When the rooster sings for the first time' or the second time, and so on. And then we're ready for the cows again."

Contrast the natural clocks of Burundi to the clock time scheduling that prevails in the dominant Anglo culture of the United

States. Our watches dictate when it is time to work and when to play; when each encounter must begin and end.

Even biological events are typically scheduled by the clock. It is normal to talk about it being "too early to go to sleep" or "not yet dinner time," or too late to take a nap or eat a snack. The hour on the clock, rather than the signal from our bodies, usually dictates when it is time to begin and stop. We learn these habits at a very early age. A newborn is fully capable of recognizing when he or she is hungry or sleepy. But it is not long before parents either adjust their baby's routine to fit their own or, in response to whatever may be the prevailing cultural standards (often defined by popular Dr. Spock–type advice manuals), train the child to eat and sleep to more "healthy" rhythms. The baby then learns when to be hungry and when to be sleepy.

As adults, some people are particularly susceptible to the control of the clock. Several years ago, in a series of classic studies, social psychologist Stanley Schachter and his colleagues observed the eating behaviors of obese and normal-weight people. Schachter theorized that a major factor in obesity is a tendency for eating to be governed by external cues from the surrounding environment. People of normal weight, he believed, are more responsive to their internal hunger pangs. One powerful external cue, Schachter hypothesized, is the clock.

To test his theory, Columbia University dormitory students were brought into a room in which the experimenters had doctored the clocks so that some subjects thought it was earlier than their usual dinner time and others thought it was later than their usual dinner time. Participants were told to help themselves to a bowl of crackers in front of them. As Schachter predicted, the obese people ate more crackers when they thought it was after their dinner time than when they were made to believe that it was not yet time for dinner. The time on the clock had no bearing on how many crackers the normal-weight subjects ate. They ate when they were hungry. The overweight people ate when the time on the clock said it was appropriate.[3] As my over-three-hundred-pound uncle replied when I once asked him if he was hungry, "I haven't been hungry in 45 years."

IS TIME MONEY?

When the clock predominates, time becomes a valuable commodity. Clock time cultures take for granted the reality of time as fixed, linear, and measurable. As Ben Franklin once advised, "Remember that time is money." But to event time cultures, for whom time is considerably more flexible and ambiguous, time and money are very separate entities.

The clash between these attitudes can be jarring. When, on my sabbatical trip, I moved out of my hot tub/breakfast/Buddha routine and made a trip to the Taj Mahal, for example, the most frequent comments I heard spoken by first-world visitors referred to the amount of work that went into the building—variations on the question, "How long must that have taken?" Perhaps the second most frequent comment I heard from tourists in India went something like: "That embroidery must have taken forever. Can you imagine how much that would cost back home?" Finding bargains on foreigners' time is, in fact, a favorite vacation activity of many Westerners. But these comments wouldn't mean much to the Indian artist who spent months embroidering a fabric or to their ancestors who'd built the Taj Mahal. When event time predominates, the economic model of clock time makes little sense. Time and money are independent entities. You need to give time to time, as they say in Mexico.

In my travels in South America and Asia I have repeatedly been confused, and sometimes even harassed, by comments such as: "Unlike you Americans, time is not money for us." My usual response is something like: "But our time is all we have. It's our most valuable, our only really valuable, possession. How can you waste it like that?" Their typical retort—usually in a less frantic tone than my own—begins with unqualified agreement that time is, indeed, our most valuable commodity. But it is for exactly this reason, event timers argue, that time shouldn't be wasted by carving it into inorganic monetary units.

Burundi again provides a case in point. "Central Africans," Salvatore Niyonzima says, "generally disregard the fact that time is always money. When I want the time to wait for me, it does. And when I don't want to do something today—for any reason, whatever reason—I can just decide to do it tomorrow and it will be as good as today. If I lose some time I'm not losing something very important because, after all, I have so much of it."

Jean Traore, an exchange student from Burkina Faso in Eastern Africa, finds the concept of "wasting time" confusing. "There's no such thing as wasting time where I live," he observes. "How can you waste time? If you're not doing one thing, you're doing something else. Even if you're just talking to a friend, or sitting around, that's what you're doing." A responsible Burkina Faso citizen is expected to understand and accept this view of time, and to recognize that what is truly wasteful—sinful, to some—is to not make sufficient time available for the people in your life.

Mexico is another example. Frustrated U.S. businesspeople often complain that Mexicans are *plagued* by a lack of attachment to time. But as writer Jorge Castaneda points out, "they are simply different . . . Letting and watching time go by, being late (an hour, a day, a week), are not grievous offenses. They simply indicate a lower rung on the ladder of priorities. It is more important to see a friend of the family than to keep an appointment or to make it to work, especially when work consists of hawking wares on street corners." There is also an economic explanation: "There is a severe lack of incentives for being on time, delivering on time, or working overtime. Since most people are paid little for what they do, the prize for punctuality and formality can be meaningless: time is often not money in Mexico."

Event time and clock time are not totally unrelated. But event time encompasses considerably more than the clock. It is a product of the larger gestalt; a result of social, economic, and environmental cues, and, of course, of cultural values. Consequently, clock time and event time often constitute worlds of their own. As Jorge Castaneda observes about Mexico and the United States, "time divides our two countries as much as any other single factor."[4]

Life in industrialized society is so enmeshed with the clock that its inhabitants are often oblivious to how eccentric their temporal beliefs can appear to others. But many people in the world aren't as "civilized" as us. (Psychologist Julian Jaynes defines civilization as "the art of living in towns of such size that everyone does not know everyone else.") Even today, organic clocks like Burundi's time of the cows are often the only standard that insiders are willing to accept. For many, if not most, people in the world, living by mechanical clocks would feel as abnormal and confusing as living without a concrete schedule would to a Type A Westerner.

Anthropologists have chronicled many examples of contemporary event time cultures. Philip Bock, for example, analyzed the temporal sequence of a wake conducted by the Micmac Indians of Eastern Canada. He found that the wake can be clearly divided into gathering time, prayer time, singing time, intermission, and mealtime. But it turns out that none of these times are directly related to clock time. The mourners simply move from one time to another by mutual consensus. When do they begin and end each episode? When the time is ripe and no sooner.[5]

Robert Lauer tells of the Nuers from the Sudan, whose calendars are based on the seasonal changes in their environment. They construct their fishing dams and cattle camps, for example, in the month of *kur*. How do they know when it is *kur*? It's *kur* when they're building their dams and camps. They break camp and return to their villages in the months of *dwat*. When is it *dwat*? When people are on the move.[6] There's an old joke about an American on a whirlwind tour of Europe who is asked where he is. "If it's Tuesday," he responds, "this must be Belgium." If Nuers were asked the same question they might answer: "If it's Belgium, this must be Tuesday."

Many people use their social activities to mark time rather than the other way around. In parts of Madagascar, for example, questions about how long something takes might receive an answer like "the time of a rice-cooking" (about half an hour) or "the frying of a locust" (a quick moment). Similarly, natives of the Cross River in Nigeria have been quoted as saying "the man died in less than the time in which maize is not yet completely roasted" (less than fifteen minutes). Closer to home, not too many years ago the *New English Dictionary* included a listing for the term "pissing while"—not a particularly exact measurement, perhaps, but one with a certain cross-cultural translatability.

Most societies have some type of week, but it turns out it's not always seven days long. The Incas had a ten-day week. Their neighbors, the Muysca of Bogota, had a three-day week. Some weeks are as long as sixteen days. Often the length of the week reflects cycles of activities, rather than the other way around. For many people, their markets are the main activity requiring group coordination. The Khasis of Assam, Pitirim Sorokin reports, hold their market every eighth day. Being practical people, they've made their week eight days long and named the days of the week after the places where the main markets occur.[7]

Natives of the Andaman jungle in India are another people with little need to buy calendars. The Andamanese have constructed a complex annual calendar built around the sequence of dominant smells of trees and flowers in their environment. When they want to check the time of year, the Andamanese simply smell the odors outside their door.[8]

The monks in Burma have developed a foolproof alarm clock. They know it is time to rise at daybreak "when there is light enough to see the veins in their hand."[9]

There are groups who, even though they have wristwatches, prefer to measure time imprecisely. The anthropologist Douglas Raybeck, for example, has studied the Kelantese peasants of the Malay Peninsula, a group he refers to as the "coconut-clock" people. The Kelantese approach to time is typified by their coconut clocks—an invention they use as a timer for sporting competitions. This clock consists of a half coconut shell with a small hole in its center that sits in a pail of water. Intervals are measured by the time it takes the shell to fill with water and then sink-usually about three to five minutes. The Kelantese recognize that the clock is inexact, but they choose it over the wristwatches they own.[10]

Some people don't even have a single-word equivalent of "time." E. R. Leach has studied the Kachin people of North Burma. The Kachin use the word *ahkying* to refer to the time of the clock. The word *na* refers to a long time, and *tawng* to a short time. The word *ta* refers to springtime and *asak* to the time of a person's life. A Kachin wouldn't regard any of these words as synonymous with another. Whereas time for most Westerners is treated as an objective entity—it is a noun in the English language—the Kachin words for time are treated more like adverbs. Time has no tangible reality for the Kachin.[11]

Many North American Indian cultures also treat time only indirectly in their language. The Sioux, for example, have no single word in their language for "time," "late," or "waiting." The Hopi, observes Edward Hall, have no verb tenses for past, present, and future. Like the Kachin people, the Hopi treat temporal concepts more like adverbs than nouns. When discussing the seasons, for example, "the Hopi cannot talk about summer being hot, because summer is the quality hot, just as an apple has the quality red," Hall reports. "Summer and hot are the same! Summer is a *condition*: hot." It is difficult for the Kachin and the Hopi to conceive of time as a quantity. Certainly, it is not equated with money and the clock. Time only exists in the eternal present.

Many Mediterranean Arab cultures define only three sets of time: no time at all, now (which is of varying duration), and forever (too long). As a result, American businessmen have often encountered frustrating communication breakdowns when trying to get Arabs to distinguish between different waiting periods—between, say, a long time and a very long time.[12]

I ran into similar dictionary problems once when trying to translate a time survey into Spanish for a Mexican sample. Three of my original English questions asked people when they would "expect" a person to arrive for a certain appointment, what time they "hoped" that person would arrive, and how long they would "wait" for them to arrive. It turns out that the three English verbs "to expect," "to hope," and "to wait" all translate into the single Spanish verb "*esperar.*" (The same verb is used in Portuguese). I eventually had to use roundabout terms to get the distinctions across.

There is an old Yiddish proverb that says, "It's good to hope, it's the waiting that spoils it." Compare this to a culture whose language does not routinely distinguish between expecting, hoping, and waiting, and you have a pretty clear picture of how the latter feels about the clock. At first, I was frustrated by the inability to translate my questionnaires. Later, though, I came to see that my translation failures were telling me as much about Latin American concepts of time as were their responses to my formal questions. The silent and verbal languages of time feed upon each other.

KEEPING EVERYTHING FROM HAPPENING AT ONCE

The primary function of clock time, it may be argued, is to prevent simultaneously occurring events from running into one another. "Time is nature's way of keeping everything from happening at once," observes a contemporary item of graffiti. The more complex our network of activities, the greater the need to formalize scheduling. A shared commitment to abide by clock time serves to coordinate traffic. The Khasis and Nuers are able to avoid governance by the clock because the demands on their time are relatively distinct and uncomplicated.

But we don't have to cross continents to see groups still operating on event time. Even in clock-time-dominated cultures, there are people whose temporal demands more closely resemble the sparsity of Asian villagers than that of the surrounding clock-coordinated society. In these subcultures, life takes on the cadence of event time.

Alex Gonzalez, a fellow social psychologist raised in a Mexican-American barrio in Los Angeles, has described the attitude toward time among his childhood friends who remain in his old neighborhood. Many of these people are unemployed, have little prospect of employment, and, he observes, almost no future time perspective. His old neighborhood, Gonzalez says, is filled with people who congregate loosely each day and wait for something to capture their interest. Their problem is not so much finding time for their activities as it is to find activities to fill their time. They stay with the event until, by mutual consent, it feels like time to move on. Time is flat. Watches are mostly ornaments and symbols of status. They're rarely for telling time.[13]

How would these people react if you gave them a Day Runner? Probably like Jonathan Swifts's Lilliputians did to Gulliver, who looked at his watch before doing anything. He called it his oracle. The Lilliputians he met in his travels decided that Gulliver's watch must be his God. In other words, they thought he was crazy.

THE ADVANTAGE OF TEMPORAL FLEXIBILITY

Clock time cultures tend to be less flexible in how they schedule activities. They are more likely to be what anthropologist Edward Hall calls monochronic or M-time schedulers: people who like to focus on one activity at a time. Event time people, on the other hand, tend to prefer polychronic or P-time scheduling: doing several things at once.[14] M-time people like to work from start to finish in linear sequence: the first task is begun and completed before turning to another, which is then begun and completed. In polychronic time, however, one project goes on until there is an inclination or inspiration to turn to another, which may lead to an idea for another, then back to the first, with intermittent and unpredictable pauses and resumptions of one task or another. Progress on P-time occurs a little at a time on each task.

P-time cultures are characterized by a strong involvement with people. They emphasize the completion of human transactions rather than keeping to schedules. Two Burundians deep in conversation, for example, will typically choose to arrive late for their next appointment rather than cut into the flow of their discussion. Both would be insulted, in fact, if their partner were to abruptly terminate the conversation before it came to a spontaneous conclusion. "If you value people," Hall explains about the sensibility of P-time cultures, "you must hear them out and cannot cut them off simply because of a schedule."

P-time and M-time don't mix well. Allen Bluedorn, a professor of management at the University of Missouri, and his colleagues have found that M-time individuals are happier and more productive in M-time organizations while polychronic people do better in polychronic ones. These findings are applicable not only to foreign cultures, but also to different organizational cultures in the United States.[15]

Both M-time and clock time thinking tend to be concentrated in achievement-oriented, industrialized societies like the United States. P-time and event time are more common in third-world economies. In general, people who live on P-time are less productive—by Western economic standards, at least—than are M-time people. But there are occasions when polychronicity is not only more people-oriented, but also more productive. Rigid adherence to schedules can cut things short just when they are beginning to move forward. And as the invention of the word processor has taught even the most rigid of M-time people, working in nonlinear progression, spontaneously shifting attention from one section of a project to another, making connections from back to front as well as vice-versa, can be both liberating and productive.

The most fruitful approach of all, however, is one that moves flexibly between the worlds of P-time and M-time, event time and clock time, as suits the situation. Some of the newer entrants into the economics of industrialization have managed monetary success without wholesale sacrifice of their traditional commitment to social obligations. Once again, the Japanese, with their blend of traditional Eastern and modern Western cultures, provide a noteworthy example.

A few years ago, I received a letter from Kiyoshi Yoneda, a businessman from Tokyo who has spent more than five years living in the West. My research on cross-national differences in the pace of life, which found that the Japanese had the fastest pace of life in the world, had just been reported in the international press. Mr. Yoneda wrote because he was concerned (with good reason, I might add) about the superficiality of my understanding of Japanese attitudes toward time. He wanted me to understand that the Japanese may be fast, but that doesn't mean that they treat the clock with the same reverence as people in the West.

Meetings in Japan, he pointed out, start less punctually and end much more "sluggishly" than they do in the United States. "In the Japanese company I work for," he wrote, "meetings go all the way until some agreement is made, or until everybody is tired; and the end is not sharply predefined by a scheduled time. The agreement is often not clearly stated. Perhaps in order to compensate [for] the unpredictability of the closing time of a meeting, you are not blamed if you [go] away before the meeting is over. Also, it's quite all right to sleep during the meeting. For instance, if you are an engineer and not interested in the money-counting aspects of a project, nobody expects you to stay wide awake paying attention to discussions concerning details of accounting. You may fall asleep, do your reading or writing, or stand up to get some coffee or tea."

Monochronic and polychronic organizations each have their weaknesses. Monochronic systems are prone to undervaluing the humanity of their members. Polychronic ones tend toward unproductive chaos. It would seem that the most healthy approach to P-time and M-time is to hone skills for both, and to execute mixtures of each to suit the situation. The Japanese blend offers one provocative example of how people take control of their time, rather than the other way around.

MORE TIME WARS

Because cultural norms are so widely shared by the surrounding society, people often forget that their own rules are arbitrary. It is easy to confuse cultural normalcy with ethnocentric superiority. When people of different cultures interact, the potential for misunderstanding exists on many levels. For example, members of Arab and Latin cultures usually stand much closer when they're speaking to people than we do in the United States, a fact we frequently misinterpret as aggression or disrespect. Similarly, we often misconstrue the intentions of people with temporal customs different from our own. Such are the difficulties of communicating the silent languages of culture.

Nearly every traveler has experienced these blunders, in the forms of their own misunderstanding of the motives of the surrounding culture as well as others' misinterpretations of theirs. A particularly frequent source of mishaps involves clashes between clock time and event time. Fortunately, most of our stumblings are limited to unpleasant miscommunications. When misunderstandings occur at a higher level, however, they can be serious business.

An example of this occurred in 1985, when a group of Shiite Muslim terrorists hijacked a TWA jetliner and held 40 Americans hostage, demanding that Israel release 764 Lebanese Shiite prisoners being held in prison. Shortly after, the terrorists handed the American hostages over to Shiite Muslim leaders, who assured everybody that nothing would happen if the Israelis met their demands.

At one point during the delicate negotiations Ghassan Sablini, the number-three man in the Shiite militia Amal (who had assumed the role of militant authorities), announced that the hostages would be handed back to the hijackers in two days if no action was taken on their demand that Israel release its Shiite prisoners. This created a very dangerous situation. The U.S. negotiators knew that neither they nor the Israelis could submit to these terrorist demands without working out a face-saving compromise. But by setting a limit of "two days" the Shiite leaders made a compromise unlikely and had elevated the crisis to a very dangerous level. People held their breath. At the last minute, however, Sablini was made to understand how his statement was being interpreted. To everyone's relief, he explained: "We said a couple of days but we were not necessarily specifying 48 hours."[16]

Forty deaths and a possible war were nearly caused by a miscommunication over the meaning of the word "day." To the U.S. negotiators, the word referred to a technical aspect of time: 24 hours. For the Muslim leader, a day was merely a figure of speech meaning "a while." The U.S. negotiators were thinking on clock time. Sablini was on event time.

FIVE

TIME AND POWER

The Rules of the Waiting Game

> It's odd how people waiting for you stand out far less
> clearly than people you are waiting for.
> JEAN GIRAUDOUX, *Tiger at the Gates,* Act 1

The classic serviceman's refrain of "Hurry up and wait" has become a truism of our times. We wait for buses and elevators, in stores and in traffic, for waiters and clerks. We line up for tickets for trains that leave late, and for appointments with doctors who leave us cooling our heels.

Waiting is unpleasant. "Half the agony of living is waiting," was the feeling of the writer Alexander Rose. While psychologists would have difficultly so precisely quantifying the pain caused by waiting, we have more than enough evidence that its effects are often deleterious. Studies have shown reactions ranging from mild frustration to ulcers to heightened susceptibility to death from coronary heart disease. Researcher Edgar Osuna has gone so far as to specify a direct mathematical relationship between waiting time and stress and anxiety.[1]

If you believe time is money, waiting is also expensive. Many scholars believe that wasted time was one of the primary patholo-gies causing the eventual breakdown of the Soviet Union. The economist Y. Orlov, for example, estimated that about 30 billion hours per year were being wasted waiting during shopping activi-ties alone: the equivalent of a year's work for 15 million people. In Moscow alone, another study estimated, more than 20 million man-hours were spent queuing just for the payment of rent and utilities.[2]

In the United States, a 1988 series of field studies conducted by researchers for the consulting firm Priority Management Pitts-burgh Inc. found that the average American spent the equivalent of five years of life standing in line, six months sitting at traffic lights and two years trying to return phone calls.[3] Although waiting in the United States may not be quite the national pastime that it is in the former Soviet republics, our time consciousness causes the loss to feel no less wasteful.

It is not surprising, given our accelerating population growth and dwindling resources, that the lines are growing longer and the problems more serious. But there is more to waiting than frustra-tion and costs. Of equal importance to social scientists are the dy-namics of the waiting game, which provide a precious opportunity for understanding the fundamental workings of culture. The rules and principles governing waiting—who waits up front, who waits in back, and who does not wait at all—are yielding precious infor-mation. These rules are part of the silent language of culture. They're written nowhere, but the messages they carry often speak louder than words.[4]

Rule One
Time is money.

In most of the world, this is the basic rule from which all the oth-ers are derived. Workers are paid by the hour, lawyers charge by the minute, and advertising is sold by the second. Through a curi-ous intellectual exercise, the civilized mind has reduced time—that most obscure and abstract of all intangibles—to the most objective of all quantities: money. With time and things on the same value scale, it is now possible to figure out how many working hours equal the price of a color television.

Not only do we sell our time for money, but there is a market for buying our own time back. "The hot new family commodity is 'off time,'" according to Heloise, who writes a syndicated column on household hints. "If I can give them another 20 minutes, even if it costs them $4 in dry cleaning, then I'm successful," she observes about her role.[5]

Public opinion polls now routinely ask "overworked" Americans to evaluate how much money they would trade for more time to themselves. In the 1991 national survey of time values conducted by the Hilton Hotel Corporation, two-thirds of all respondents said they would take salary cuts in exchange for getting time off from work. This willingness to exchange money for time was consistent across gender, age groups, educational background, economic status, and number of children.[6] "Time or money?" becomes the question, as if the two were interchangeable forms of the same elastic entity, like water and ice, or cash and checks.

It is a very different story in cultures that do not play by the time-is-money rule. Once, in Kathmandu, Nepal, I needed to make a phone call back to the United States. This is a country where people must turn their watches forward by exactly ten minutes when crossing the border from India. And, for God knows what reason, Kathmandu is five hours and forty minutes ahead of Greenwich mean time. I was prepared for the worst.

To beat the crowds, I went to the main phone center at 7:30 in the morning. At the international window, I waited ten minutes for the clerk to approach. It would take a few minutes to get through to an international operator, he said. I sat down, took out a book, and waited.

And waited. By 9:30 I hadn't heard anyone's name called. Nor did I see the original clerk. To check on my status, I went back to the window, which by now had a long line in front of it. I waited for 25 minutes. When it was my turn the new clerk told me that the phone lines were busy and I would just have to be patient.

By noon my name still hadn't been called. The new clerk was also gone. I went back up to the window. Following a 45-minute wait this time—I'd hit the lunchtime rush—the newest clerk had difficulty locating my request. After several minutes, he discovered it in the "inactive" pile. It showed that the caller (me) had given up and left.

I sat back down and bought some lunch from a walking restaurant who was now circulating through the waiting area. Somehow, seeing a man whose livelihood depended on selling full-course meals to people waiting to make phone calls did nothing for my confidence in the phone company. I finished my book. Finally, at almost 3:00, the clerk announced to us all that the overseas lines all had gone dead. Would we please return tomorrow?

The next morning, after nearly two hours, my name was called. I ran forward. "I just wanted to verify the number you're calling," the clerk said. I confirmed that he had the digits right.

"Please sit down and wait," he said. "I'll call you when I get the operator."

For some reason this exchange didn't anger me. In some deranged way, in fact, I felt I'd taken a giant step forward. Before this small piece of recognition I felt more than a little like the protagonist in Kafka's *The Trial*, who knows neither what he is charged with, why he is there, what he is waiting for, or how long he will be forced to wait for it. I returned to the waiting area feeling a little above my compatriots, who had still not left their seats. Several wanted to know what had gone on up at the window. I begrudgingly revealed the specifics of my personal audience with the clerk. I was a big shot.

I sat down and waited some more. And before long I heard it. Perhaps it sounded like a whisper to those who didn't grasp its significance, but to my ears it was a regal announcement: "Looween, your call is waiting in booth number two," the clerk called out.

Feeling the eyes of my less fortunate fellow waiters on me, I walked slowly to the desk, trying to exude the dignity expected of a man in my position—a man whose name had been called twice within one hour by the clerk at the Kathmandu telephone office. I thought about how I would soon be asked to address my fellow waiters about my long, lonely journey to booth number two; how I had never given up hope; that sometime soon it would be they, not Mr. Looween, who made this triumphant walk.

I entered the booth. The clerk asked if I was ready. Hey, does a Buddhist say *om mani padme hum*? Then I heard the lovely voice of the overseas operator. In perfect English she asked, "Is this Mr. Lou Green?" I improvised.

"Yes, this is Mr. Green," I answered.

"Go ahead, please. Your party is on the line."

"Hello, Beverly," I shouted. "It's Bob, Bob Le Green. I'm in Nepal. In phone booth number two." From the other end I heard something like the noise that my dog used to make when someone stepped on her tail. Then silence. I yelled for the clerk. He said there must be a problem with the connection. He'd have to call the overseas operator back. "Have a seat. It will be a few minutes."

I looked toward the waiting area. The same faces were there. Every last motley one of them. It is difficult to express just how distasteful the thought of sitting again among my lowly colleagues-in-waiting felt. I left the building. Back to base camp, a night's sleep, and a new day.

The next morning was kinder. I waited perhaps twenty minutes before the clerk (the second one from the first day) announced to us all that there would be no international calls today: "The King has all the lines tied up." As another foreigner sitting next to me commented, "That's certainly one hell of a good excuse."

It wasn't until the fourth day that my call finally went through. But what I was struck by even more than the delay was how little my fellow waiters shared my suffering. Waiting, it seems, is indigenous to so much of the lives of most Nepalese that having to wait hours to make a long distance call is neither unexpected nor particularly taxing.

But the Nepalese phone company might note that the "time is money" rule can only be bent so far. They might consider what happened to their colleagues across the border in India.

The phone service in India is so slow that some businesses keep a boy on the payroll just to wait for a dial tone to put a call through. If the desired number is reached, he keeps it open under the possibility that someone might need it later in the day. A few years ago, the phone company crossed the line. P. C. Sethi, a member of Parliament, had spent most of a day trying to phone Bombay. When the call still hadn't been completed after several hours, Sethi took a gun, got an armed escort and stormed the telephone exchange. The English language newspapers carried front page headlines like "P. C. Sethi Goes Berserk." Monopoly or not, Sethi had decided that the company had broken the rules by keeping a man of his importance waiting so long.

Armed assault is admittedly an extreme reaction to frustration, but in this case it received considerable sympathy. As the manager of one Indian company remarked: "This [phone] system can only work at the point of a gun. I appreciate what [Sethi] did. I am going to write a letter telling him I agree with him."[7]

Rule Two
The law of supply and demand regulates the line.

Where time is money, it is governed by the usual rules of economics. We wait for what we value. The greater the demand, and the scarcer the supply, the longer the line. So people wait in lines to hear popular performers, in traffic to reach favorite beaches, and in offices to consult prestigious lawyers.

When demand overwhelms supply, waiting time may exceed the original value of the product itself. In these cases, the time we wait literally becomes the cost of the product. In Communist Poland, for example, I once watched people wait more than two hours for the privilege of buying a pair of shoes (and "no time for trying on, please"). As soon as they left the store, many lucky customers turned around and offered their purchases at black-market prices. The resale price, I learned, was simply calculated by how long the original buyer had to wait in line. The quality of the shoes, in a town where no alternatives were available, was beside the point. And anyone who has ever entered into negotiations with ticket scalpers here in the United States knows that it is not only the Eastern Europeans who equate waiting time with value.

People sometimes gauge the value of time as if it were a commodity on the stock market. In a *Time* magazine cover story entitled "How America has run out of time," Nancy Gibbs observed:

> There was once a time when time was money. Both could be wasted or both well spent, but in the end gold was the richer prize. As with almost any commodity, however, value depends on scarcity. And these are the days of the time famine . . . If all this continues, time could end up being to the '90s what money was to the '80s. In fact, for the callow yuppies of Wall Street, with their abundant salaries and meager freedom, leisure time is the one thing they find hard to buy.

Louis Harris, whose polls show a 37 percent decrease in Americans' leisure time over the past twenty years, asserts that, "Time may have become the most precious commodity in the land.[8]

As the price of time goes up, the rules governing its distribution become more significant. The waiting game becomes a high-stakes affair.

Rule Three
We value what we wait for.

> Make 'em laugh, make 'em cry, but above all, make 'em wait.
> BILL SMETHURST, soap opera producer

This is a psychological corollary that helps rationalize the wounds of the wait. It turns out, oddly enough, that we actually *believe* the shoes are more valuable when the lines are long. There are at least two reasons for this.

First, there is the psychological need to justify our expense—

our time, in this case. This is known in social psychology as the law of cognitive dissonance: We are motivated to find—or, when necessary, to fabricate—an explanation for behaviors that would otherwise make us feel foolish. Faced with the alternatives of believing "What a jerk I was to waste my valuable time for such a crappy pair of shoes" versus "These treasures were sure worth the wait," most healthy people opt for the latter.

When something is too easily available, in fact, people often don't want it. Who likes to eat in an empty restaurant? We tell ourselves we chose the restaurant with the long line because it probably serves better food. But the wait itself is an important part of the attraction. Somehow the meal doesn't seem quite as appetizing without a crowd.

Second, there is a human motive to value that which is least available. The social psychologist Robert Cialdini spent three years infiltrating various organizations to observe the techniques used by influence professionals—people who make a living by getting others to comply with their wishes. One of the most effective techniques, he learned, takes advantage of what is called the scarcity principle: the less available an opportunity, the more precious it seems. The clever influence professional often exploits tentative buyers by manipulating them to believe that they will face a long wait if they don't act immediately. Cialdini recalls:

The tactic was played to perfection in one appliance store I investigated where 30 to 50 percent of the stock was regularly listed on sale. Suppose a couple in the store seemed, from a distance, to be moderately interested in a certain sale item . . . a salesperson might approach and say, "I see you're interested in this model here, and I can understand why; it's a great machine at a great price. But, unfortunately, I sold it to another couple not more than 20 minutes ago. And, if I'm not mistaken, it was the last one we had."

The customers' disappointment registers unmistakably. Because of its lost availability, the appliance suddenly becomes more attractive. Typically, one of the customers asks if there is any chance that an unsold model still exists in the store's back room or warehouse or other location. "Well," the salesperson allows, "that is possible, and I'd be willing to check. But do I understand that this is the model you want and if I can get it for you at this price, you'll take it?" Therein lies the beauty of the technique. In accord with the scarcity principle, the customers are asked to commit to buying the appliance when it looks least available and therefore most desirable. Many customers do agree to a purchase at this singularly vulnerable time.[9]

It has also been found that customers who then learn the desired model is back in good supply—that nobody will need to wait—often find it less attractive again. The longer the line, the better it looks.

Waiting, then, originates out of limited resources. But the laws of economics are only the beginning. How these resources are distributed forms the real heart of the waiting game. If you look more closely you see the essence of status, power, and self-worth.

Rule Four
Status dictates who waits.

The more important we are, the greater the demand for our time. And since time is limited, its value increases with our perceived importance. Like any valuable commodity, important people's time must be protected. This leads to two corollaries of this rule. Important people are usually seen by appointment only; and while those of higher status are allowed to make people below them wait, the reverse is strictly prohibited. As long as those on top offer something of value—be it a product, a service, access to valuable resources, or simply the pleasure of contact—these rules are legitimized.

Status distinctions in the waiting game are sometimes remarkably precise. In universities like my own, for example, there used to be an unwritten rule that students must wait 10 minutes for an assistant professor who is late for class, 20 minutes for an associate professor and 30 minutes for a full professor. There is a world of difference when the roles are reversed. In a study by psychologists James Halpern and Kathryn Isaacs, groups of students and professors were asked how long they would wait for both a student and for a professor who was late for an appointment. All respondents said they would wait significantly longer for a professor than for a student. And students, compared to professors, said they would wait a great deal longer for any appointment—whether it was with a student or with a professor.[10] These standards proclaim our relative worth: I'm permitted to waste my students' time, with no explanation required, but they're prohibited from encroaching on mine.

This relationship sometimes leads to extraordinary encounters. Experienced instructors are pathetically aware that nearly every student's favorite lecture is no lecture at all. When I guiltily announce that a class will be canceled I'm *always* received with exhilarated cries of "all right!," hand-clapping, and even Dixie victory howls—none of which I have heard after even my very best lectures. And this is from students who like me.

What makes all this so peculiar is that those students pay my salary. From a monetary standpoint, they own my time, rather than the reverse. But along with my professorship, I've been granted control of their grades, which to some degree determine their future. In the end, our rules of waiting make it clear just who is running the show.

In some countries, making others wait is the essence of status. In a survey in Brazil, my colleagues and I asked people how much they thought punctuality for appointments was tied to success. To my surprise, Brazilians rated people who are always late for appointments as most successful and punctual people as least successful. Our data also showed that Brazilians rated a person who was always late for appointments as more relaxed, happy, and likeable—all of which tend to be associated with being successful.[11]

These answers threw me at first. Even in a country of seemingly infinite temporal tolerance, this appeared to be going overboard. It is one thing to be flexible, but another to believe that not getting there on time actually pays off. Here I was hoping to break through the simple *amanhã* stereotype and instead felt like I had landed in the middle of an old Carmen Miranda movie. But I was missing the point.

Brazilians rate people who are always late for appointments as most successful because this is a fact. Important people keep their underlings waiting. It is not so much that lack of punctuality causes success as that it is a result of success. Lack of promptness is a badge of achievement. It's part of the outfit, like nice shoes. In the United States we resent it when powerful people, such as doctors, keep us waiting. But the Brazilians we questioned didn't resent having to wait any more than they resented earning less money than their superiors. They envied it. Someday they, too, hoped to be successful enough to own a nice home and an expensive car and to keep others waiting.

There is a practice in many Arab cultures whereby a young woman who is caught being intimate with a man she is not married to is sometimes murdered by her brothers. To Westerners, this is uncivilized behavior. But the brother is committed to protecting the role of an important institution—the family—in the social pattern. It is his responsibility. The sister is a sacred, inviolable link between families and it is imperative to the survival of the social order that she remain above reproach. The temporal behavior of important Brazilians must, similarly, be understood as part of a larger pattern. The rule is to wait for he who holds the keys. And, in Brazil at least, no whining allowed.

Sometimes the status rule can lead to amusing power struggles, as the writer E. B. White observed many years ago, in his description of an "Impasse in the Business World":

> While waiting in the antechamber of a business firm, where we had gone to seek our fortune, we overheard through a thin partition a brigadier general of industry trying to estab-

lish telephone communication with another brigadier general, and they reached, these two men, what seemed to us a most healthy impasse. The phone rang in Mr. Auchincloss's office, and we heard Mr. Auchincloss's secretary take the call. It was Mr. Birstein's secretary, saying that Mr. Birstein would like to speak to Mr. Auchincloss. "All right, put him on," said Mr. Auchincloss's well-drilled secretary, "and I'll give him Mr. Auchincloss." "No," the other girl apparently replied, "you put Mr. Auchincloss on, and I'll give him Mr. Birstein." "Not at all," countered the girl behind the partition. "I wouldn't dream of keeping Mr. Auchincloss waiting."

This battle of the Titans, conducted by their lieutenants to determine which Titan's time was the more valuable, raged for five or ten minutes, during which interval the Titans themselves were presumably just sitting around picking their teeth. Finally one of the girls gave in, or was overpowered, but it might easily have ended in a draw. As we sat there ripening in the antechamber, this momentary paralysis of industry seemed rich in promise of a better day to come—a day when true equality enters the business life, and nobody can speak to anybody because all are equally busy.[12]

Rule Five

The longer people will wait for you, the greater your status.

Do try and see the thing primarily in its simplicity,
the waiting, the not knowing why, or where, or when,
or for what.

SAMUEL BECKETT

The inverse of rule four—that status dictates who must wait—is also true: Your position in the waiting hierarchy often determines your importance. As was the case for the resale price of the Polish shoes, a longer line for people makes them more important and expensive. The value of financial consultants, attorneys, or performers is enhanced by the simple fact that they are booked well in advance. This leads to even greater demand for their time, and so the cycle continues.

In large companies, the social boundaries are sometimes reflected in the architectural organization of the building. In a literal sense, the higher up you go, the longer the wait. Barry Schwartz in his book *Queuing and Waiting* relates one such experience:

> Low down on the scale are the men you can walk right up to. They are usually behind a counter waiting to serve you on the main floor, or at least on the lower floors. As you go up the bureaucracy you find people on the higher floors and in offices: first bull pens, then private offices, then private offices with secretaries—increasing with each step the inaccessibility and therefore the necessity for appointments and the opportunity to keep people waiting. Recently, for example, I had an experience with a credit card company. First, I went to the first floor where I gave my complaint to the girl at the desk. She couldn't help me and sent me to the eighth floor to talk to someone in a bullpen. He came out, after a suitable waiting time, to discuss my problem in the reception room. I thought that if I were to straighten this matter out I was going to have to find a vice-president in charge of something, who would keep me waiting the rest of the day. I didn't have time to wait so I took my chances with said clerk, who, of course, didn't come through. I'm still waiting for the time when I have an afternoon to waste to go back and find that vice-president to get my account straightened out.[13]

When the server's worth becomes psychologically enhanced by cognitive dissonance and the law of scarcity (rule three), the least accessible people are sometimes elevated to saviorlike dimensions. An extreme example of this occurs in Samuel Beckett's play *Waiting for Godot*. Godot is a server who does not serve. His very value derives from the fact that he is waited for.

For the people doing the waiting, on the other hand, there's nothing like a long delay to put them in their place. They needn't be reminded that the original meaning of the term "waiter"—as the French term "*attendant*" makes clear—was one who served the whims of his superior.

The position of the waiter becomes particularly obvious when we are asked to wait during the appointment itself—for example, when someone we have been trying hard to reach deigns to see us and then answers a phone call while we sit there. How humiliating when he or she turns back to the business we have waited weeks to discuss, their train of thought is lost, and we're told, apologetically, that our matter will have to wait for another time. Why not just say it to our face: "You're less important than me than the person I just spoke to, and whatever else might come up the rest of this hour."

Rule Six
Money buys a place in front.

Not only may important people make their subordinates wait, but there is a privileged class who are nearly immune from waiting. They get special services that spare their precious time, which they then use to earn more money to pay for special services.

The elite can afford, for example, to shop in stores where salespeople meet customers at the door or, if they are even more fortunate, to send others to shop for them. Should they want tickets for a sold-out concert they call a ticket broker. They don't even wait at the bank for their money matters. When you have a large enough account, the bank comes to you.

It is a very different story for those on the bottom. Even when waiting for the very same services, the line is longest for people without resources. In a study by Barry Schwartz,[14] for example, a national sample of Americans were asked how long they typically had to wait in a doctor's office. African-American people—who tended to be of lower socioeconomic standing—reported waiting longer than whites. And the lower their socioeconomic status, the longer their wait: 36 percent of high-status whites reported waiting 30 minutes or more, compared to 50 percent of high-status blacks,

51 percent of low-status whites and 69 percent of low-status blacks.

In some countries, there are people whose sole function is to stand in lines for the well-off—to be something of a temporal wet-nurse. In Mexico, whose bureaucracy makes that of the United States look like a well-oiled machine, life is clogged with endless *tramites*, or bureaucratic procedures. Renewing a driver's license, for example, can take a full day of waiting in lines. For those with money, however, there are people known as *gestores*—or, in less formal terms, *coyotes*—who hire themselves out as substitute wait-ers. A number of gestores can be found outside government ser-vice agencies. Some even have their own offices. Clients simply hand over their information—the gestor usually carries a full sup-ply of necessary blank forms—and arrange a time to pick up the completed papers.

The negotiated price is based on the amount of time the cus-tomer will be saved. One of the responsibilities of an efficient gestor is to speed up the bureaucratic process, usually by making payoffs to the appropriate people. As a result, the time the client saves—and pays for—may be more than what the gestor loses. Even though the work of the gestor is embedded in graft, there is noth-ing sordid about the profession's image. Gestores are seen as useful intermediaries. They provide a necessary and important service.

For some *tramites* in Mexico, it is possible to literally hire bodies to take one's place in line. Obtaining a visa, for example, often re-quires queuing up outside the consulate the night before in order to secure a tenable position when the doors open in the morning. Gathered in front of the building, however, are people who will hire themselves out to stand in your place overnight. For the right price, entire families will wait in place of your own.

In Brazil, there are more highly trained professional waiters known as *despachantes*. These are para-professionals who serve as intermediaries between well-off citizens and the endless bureau-cratic red tape. To give an idea of the extent of the paper-pushing in Brazil, during the twelve months I lived in the country the gov-ernment required that I obtain, in visas alone: an entry visa on arrival, temporary exit visas on each of the four occasions I took side trips to neighboring countries, temporary entry visas to be allowed back after each of these four trips, and an exit visa to permit me to leave Brazil at the end of my stay. In all, I was to obtain a little less than one visa per month.

This red tape crossed the line of absurdity when, near the very end of my stay, I applied for my permanent exit visa, which I had just been told would be required before I could obtain my plane ticket back home. When I approached the appropriate building, I found the usual group of despachantes, mostly well-dressed and carrying professional-looking attache cases. (Despachantes are afforded more prestige by their clients than are their gestor cousins.) I asked one, whom I'd had successful dealings with in arranging for an earlier visa, how long the process would take. If I did it myself, he explained, it would be up to three weeks, for the papers had to pass through several offices in three different buildings. But I needed to leave Brazil in two weeks, I explained. "*Nao ha problema*," he answered in Portuguese. "I can take it straight to someone I know in the third building who can sign for the all the people in the first two buildings, and I'll have it back for you in two days." "But if you can just bypass the first two buildings," I asked, "what are those agencies for in the first place?" "Nothing, really," he answered. "They're just government jobs."

As we reviewed my application materials, the despachante asked for my "Carteira de Identidade: Estrangeiro Temporario" (Temporary Identification Card for Foreigners). Notification of this status had to be obtained before I was allowed to work in Brazil—which I had now been doing for eleven months—and was a prerequisite to obtaining my original entry visa. I explained that I had applied and received approval for this temporary foreigner status before entering Brazil, but had never received the actual card. Technically then, he pointed out, I had no proof that I was eligible to enter Brazil, so I couldn't be granted approval to leave. I would need to get my entry papers before I could obtain my exit visa. The despachante was nonplussed by this added twist. "These foreigner entry papers have to go to more than one city, and that can mean close to a year. I'll have to work on that one, too."

He said he could do it all in a week. I wasn't optimistic. But six days later, as I was thinking of finding a real crook to smuggle me across the border, the despachante arrived at my door with the documents. Fifty-one weeks after entering Brazil, I was granted official permission to begin my visit, which was going to end in seven days. And, at the same moment, I received approval to leave.

In the United States, too, time-selling is currently a high-growth profession. In characteristic American style, however, entrepreneurs are targeting a much wider range of activities than those attended to by the gestores and despachantes. One example is a company called "At Your Service," founded a few years ago by Glenn Partin and Richard Rogers in Winter Park, Florida. At Your Service will stand in for virtually any task customers don't want to spend their own time doing, from waiting in line to household maintenance to shopping and running errands. This business is typical of what journalist Nancy Gibbs calls "the growing number of entrepreneurs who will perform any service within their expertise, for anywhere between $25 and $50 an hour. What was once a cottage industry of people providing household services is currently a booming business in cities all across the country. Anyone who can protect a family's free time is a sure success."[15]

Even when the wealthy cannot completely avoid waiting, they suffer less than the have-nots. Whenever possible, their waiting environment is made as comfortable as possible. Milla Alihan, in his book *Corporate Etiquette*, advises:

> Wasting the time of any businessman is equivalent to robbing his wallet. Keeping him waiting is bad business practice and bad manners.
>
> [If a delay is inevitable], your secretary should explain the circumstances and ask him if he would mind waiting. She might ask him into your private office, take his coat and hat, see that he is seated and comfortable. She might ask him if he would like her to get him a magazine to read, or coffee or tea, or a soft drink while he is waiting for you.[16]

Research indicates that this suggestion is often enforced—but selectively. In one study, Barry Schwartz and his colleagues observed the treatment accorded to clients in the executive office of a mortgage company in an eastern American city. Not only was there a direct relationship between clients' status and how long they were kept waiting, but high-status clients were more than four times as likely (36 percent versus 8 percent) to be offered a beverage during their wait than were low-status clients. High-status clients were also more than twice as likely (75 percent versus 33 percent) to receive a personal escort from the waiting room to their appointment.[17]

Often the elite are assigned completely separate accommodations. In airports, for example, they have access to VIP lounges, which provide posh eating and drinking facilities and other upscale comforts—not the least of which is isolation from the larger mass of plebeian waiters relegated to sweat it out in the ingloriously named "waiting area" or at the "gate." Even the legal system offers better waiting conditions to those with resources. While awaiting trial, the wealthy can almost always raise money for their bail, which allows them to wait at home in freedom. The poor often must wait in jail. If they are sent to prison, people with resources are considerably more likely to "do time" in more comfortable facilities, epitomized by the infamous Club Feds, than are the less privileged.

Rule Seven
The more powerful control who waits.

With status and money, then, comes the ability to control time, both your own and that of others. And here we come to the guts of the waiting game: Time is power. There is no greater symbol of domination, since time is the only possession which can in no sense be replaced once it is gone. The power principle comes as a triad: First, making a person wait is an exercise in power. Second, powerful people have the capacity to make others wait. And, third, the willingness to wait acknowledges and legitimizes this power.

Serious power players, well aware of this rule, often make direct attacks on personal time. A woman high up in Bhagwan Shree Rajneesh's former religious empire told me: "The aim of Bhagwan was absolute devotion. During introductory weekends in the United States, we would begin this process by asking new recruits to make a commitment. Our first request was that they give up their watches, followed by their money and their clothes. Bhagwan knew that once he had their time—first, symbolically and, later, literally—he had *them*."

Sometimes the powerful make others wait as a way of flexing their muscles, to remind their underlings who is in charge. The medieval pope Gregory VII was said to appreciate this exercise. He once forced the Holy Roman Emperor Henry IV—who had earlier challenged his authority—to stand barefoot in the snow and ice for three days and nights before allowing him a meeting.

The Russians, for whom waiting so often takes center stage in daily life, have a particular fascination with waiting as a weapon. Aleksandr Solzhenitsyn, for example, writes in *The Cancer Ward*:

Having met the man (or telephoned him, or even specially summoned him), he might say: "Please step into my office tomorrow morning at ten." "Can't I drop in now?" the individual would be sure to ask, since he would be eager to know what he was being summoned for and get it over with. "No, not now," Rusanov would gently, but strictly admonish. He would not say that he was busy at the moment or had to go to a conference. He would on no account offer a clear, simple reason, something that could reassure the man being summoned (for that was the crux of this device). He would pronounce the words "not now" in a tone allowing many interpretations—not all of them favorable. "About what?" the employee might ask, out of boldness or inexperience. "You'll find out tomorrow," Pavel Nikolaevich would answer in a velvet voice, bypassing the tactless question. But what a long time it is until tomorrow.[18]

Military strategists are sometimes particularly adept at using time as an offensive weapon. When they play the game well, the tables can be turned on better-armed foes. Early in Lyndon Johnson's presidency, for example, he received a memo from Nicholas Katzenbach that argued (unsuccessfully) for a halt to U.S. bombing and a gradual withdrawal of U.S. troops from Viet Nam. Clark Clifford, who served in President Lyndon Johnson's inner circle, called the memo both "striking and prophetic":

> Hanoi uses time the way the Russians used terrain before Napoleon's advance on Moscow, always retreating, losing every battle, but eventually creating conditions in which the enemy can no longer function. For Napoleon it was his long supply lines and the cold Russian winter; Hanoi hopes that for us it will be the mounting dissension, impatience, and frustration caused by a protracted war without fronts or other visible signs of success. . . . Time is the crucial element at this stage of our involvement in Vietnam. Can the tortoise of progress in Vietnam stay ahead of the hare of dissent at home?[19]

My own choice for the Waiting Game Power Player of the Decade is Iraqi leader Saddam Hussein. By every objective military yardstick, the United States and its allied forces have handed Saddam a merciless string of defeats. Baghdad has absorbed relentless bombings and almost total trade embargos with scarcely a military counterattack of its own. Saddam's only defense—and, ultimately, his offense—has been his inaction and patience. After each threat Washington waits for his response, and each time Baghdad sets the tempo.

When in August 1990 the United States first summoned its influence to place an almost total trade embargo on Iraq, the headline in *USA Today* was a prototype for the future: "Now, a waiting game." After the characteristic nonresponse from Saddam, the United States was forced to escalate its military threats. As Mideast expert Ahmad Khalidi told the *New York Times*: "With every day that passes it is very clear that the longer Saddam Hussein holds out, the more you will have a groundswell of support for Iraq."[20] A U.S. military intelligence specialist commented: "It worries me that, once again, Saddam is controlling the war's tempo at little cost to himself."[21] Even after his surrender, Saddam has exerted the same temporal control. When would his troops withdraw? When would he destroy his weapons? When would inspection teams be allowed access? Always the ultimatums came from his enemies, and always they waited for Saddam's response.

In his June 1996 profile of Saddam in the *New Yorker*, T. D. Allman observed that "Today, everyone agrees that Saddam's grip on power remains total." If victory is measured by one's staying in power, Allman concludes, "not only did Saddam emerge victorious from his 'defeat' in the Gulf War but he keeps going from strength to strength." Tariq Aziz, who has served as Saddam's foreign minister and deputy prime minister, boasts in the same article: "Of course we were victorious . . . we defeated you! Iraq has not been reduced to primitiveness and we're still in control. Bush and Baker are out of power now." When asked about how long Saddam and his leadership intended to stay in control, Aziz responded, "Forever."[22] Such is the power that comes with controlling the pace of life.

Rule Eight
Waiting can be an effective instrument of control
(or, The Siddhartha move).

Saddam's case demonstrates a special application of the power rule: how waiting can in itself be a powerful act. Recall Herman Hesse's young Siddhartha, who believed: "Everyone can perform magic, everyone can reach his goal, if he can think, wait and fast." With the right attitude, waiting is a potent tool against the obstacles of life.

The key here is to move our minds from clock time to event time; to forget the hour on the clock and the notion that time is money. Siddhartha was willing to use as much time as was necessary to achieve his goals. He recognized the value of his time, but it had nothing to do with being paid by the hour. To be governed by society's clock was, to him, a waste of his most precious resource. The poet Ranier Maria Rilke put it more succinctly. Above his desk, it is said, he had written the single word "WAIT." Rilke understood that waiting is, after all, simply the gap and link between the present and the future. It is what St. Augustine called "the present of the future."

The Siddhartha approach can be extremely effective, especially for people of otherwise limited means. But, as Saddam Hussein shows, it is not confined to spiritual pursuits. It sometimes takes unattractive forms. Walter Winchell was generally considered the most powerful journalist of his time, and he was certainly the most feared. Winchell ran at high speed; he once wrote of himself, "I live the pace that kills." But it was through waiting that he achieved his greatest power. In his autobiography, Winchell wrote about all the "ingrates" who had forsaken him:

> I have forgiven, but I don't have to forget. I'm not a fighter, I'm a "waiter." I wait until I can catch an ingrate with his fly open, and then I take a picture of it.
>
> When some heel does me dirt (after I've helped him or her) I return the compliment some day. In the paper, on the air, or with a bottle of ketchup on the skull. I don't make up nasty things to write about them. I wait until they get locked up for taking dope or pimping and then I make it Public.[23]

Siddhartha he wasn't. But Winchell well understood the master's teachings about time.

The potential power of the Siddhartha principle cannot be denied. But although it is a simple and powerful card to play, its range of use can be limited. The problem is that its application often requires a reinterpretation of rule one (time is money), an act less palatable to many people than it was to Saddam Hussein. The usual outcome for many twentieth-century Siddharthas is, I'm afraid, expressed accurately in Mary Montgomery Singleton's poem:

Ah, "all things come to those who wait,"
(I say these things to make me glad),
But something answers, soft and sad,
"They come, but often come too late."

Rule Nine
Time can be given as a gift.

Waiting may be used as a self-imposed act of generosity. Most often these offerings are direct and personal: waiting for someone to heal, being a caretaker at a deathbed. But time can also be given as a gift without face-to-face contact. Perhaps the most peculiar of all incarnations of waiting is when people choose to publicly spend their time as an offering of respect to a superior.

After John Kennedy's assassination, for example, almost a quarter of a million people waited up to 10 hours in cold weather outside the Capitol Rotunda, where his body lay in state. No superior coerced their presence. They received no gratitude. These people simply chose to offer their time to their beloved leader, just as a Buddhist might place fruit at the feet of a deity. As one participant put it: "We were going to watch it on television in our room at the 'Y.' But the more we watched the more we felt we had to do something—something."[24] In a society where time is money, voluntarily waiting is, indeed, a precious offering.

The offering is a special instance of using time to demonstrate respect. Self-imposed waiting expresses deference for another person. Emily Post's recommendation for White House etiquette, for example, states:

When you are invited to the White House, you must arrive several minutes, at least, before the hour specified. No more unforgivable breach of etiquette can be made than not to be standing in the drawing room when the President makes his entry.[25]

The respect pattern also often dictates waiting for one's superior to depart first. Millicent Fenwick's own book of etiquette commands: "The two cardinal points of White House etiquette are that no guest is late and that no guest leaves before the President and his wife have gone upstairs."[26]

The "offering" of time is notable because it goes beyond any explanation of profit and gain or supply and demand—which is where our analysis of waiting began. Its sole purpose is to send a social message. This is the undistilled silent language speaking when words will not suffice.

Rule Ten
If you do break into line, do it at the rear.

Usually those who break into line in front of others must make their move close to the rear of the line to avoid trouble. Every August, for example, thousands of Australian football fans line up overnight outside the Melbourne Cricket Ground hoping to buy one of the few remaining seats for Australia's equivalent of the Super Bowl. When psychologist Leon Mann and his colleagues studied these queues during the 1960's, they were surprised to find that queue breaking was most common near the end of the lines, where the probability of obtaining a ticket was smallest. One reason for this turned out to be that the back of the queue was relatively unorganized. The late arrivers there had been together for less time, were less familiar with each other, and so were less likely to recognize and try to stop an intruder.[27]

Another reason that more queue-breaking takes place in back is that fewer people complain. Stanley Milgram and his colleagues tested people's reactions when they broke into waiting lines at railroad ticket counters, betting parlors, and other New York City locations.[28] Of all objections, they found, 73 percent came from those standing behind the point of intrusion—the people whose chances for a ticket were affected by the queue breaker. The further back the break-in, then, the fewer the objections the intruder is likely to contend with. Extrapolating these queue dynamics to the larger social order, we see yet another way that those on the bottom are losers in the waiting game.

THE INTERNATIONAL WAITING GAME

However knowledgeable people are about waiting in their own country, learning to play this intricate game in a foreign culture is tricky. The rules are often as divergent as the countries themselves.

The British, for example, pride themselves on the orderliness of their queues. Israelis, on the other hand, stubbornly resist forming distinct lines. But when Leon Mann studied Israelis at bus stops, he found that they established implicit rules, so that commuters almost invariably boarded the bus in order of arrival. This system, he observed, reflects the orderly, egalitarian nature of Israeli society, which values independence and service according to need, but rejects regimentation.[29]

Reactions to waiting are also culturally diverse. One study, for example, found that Italian queues are more likely to be characterized by lighthearted conversation and a general atmosphere of gaiety, as opposed to the irritability and impatience that typify American lines. Another study found that Catholics in America tend to be more impatient with waiting than Protestants.[30]

Because the rules of waiting are usually not made explicit, outsiders often misinterpret the message. The inevitable result is conflict. King Hassan of Morocco, for example, is a notorious late arriver whose lack of punctuality has ultimately injured his country's foreign relations. In 1981, when Queen Elizabeth II paid a call, the King kept her waiting for fifteen minutes. The Queen was not amused.

On another occasion, Hassan was one of the few influential members of royalty absent from the Charles-Diana nuptials. Because of his eminent position, he had to be asked. But the invitation was qualified by statements alluding to the high value that Anglo-Saxons place on promptness and the hope that His Majesty could manage to be on time for the ceremony. The King responded in due course that certain pressing affairs would, unfortunately, preclude his personal appearance. He sent the Crown Prince in his place.

The Moroccans still don't understand why the British were so upset by the King's lack of promptness. "The King could never have kept the Queen or anyone else waiting," one of them later said, "because the King cannot be late."[31]

Listen closely and you'll hear the silent language.

PART II

FAST, SLOW, AND THE QUALITY OF LIFE

SIX

WHERE IS LIFE FASTEST?

Dawia maintained that the Europeans were thus fa-
vored by Allah because Allah liked automobiles and
was hoping that the Europeans would bring their
cars to Heaven with them. Omar, however, said no,
that Allah loved the Europeans because the Europeans always got to their appointments on time.

JANE KRAMER, *Honor to the Bride*

There is something about poking around in foreign cultures
that compels a person to compare—to compare one culture to another, and your own life to each of theirs. In my
own case, the comparisons always seem to center around time. For
the last ten years, my dual preoccupations, traveling and social psychology, have converged on two questions: Which cultures are
fastest or slowest? And how does this cultural tempo affect the
quality of peoples' lives? My interest in these questions has been
provoked by visits to other cultures; but I have searched for answers through the systematic methods of social science.

Comparing the personalities of different cultures is a tricky
business. Labeling individuals is complex enough; how does a scientist presume to classify whole groups of people? To measure the
tempo of life with any degree of systematic objectivity requires
moving beyond anecdotal descriptions. We need to zero in on situations that are not only informative about temporal experience,

but which also have the same psychological meaning in different cultures. Developing these measures has been more difficult than I had anticipated.

One type of situation I wanted to measure, for example, was an indicator of speed in the workplace. I needed to find a naturally occurring work behavior that would be timable, easily observable, and equivalent in meaning across cultures; also, I had to be sure the workers we tested would be residents of the country. For a while, this search ran into nothing but dead ends. One possibility that was rejected, for example, was observing ticket agents for airlines. The problem, I found, was that these workers tended to come from cities, or even countries, other than the one where they were working. It was hard to know whether their work speed reflected the norms of the country they came from, the one they were in now, or simply the culture of the airline industry.

Another failed possibility was timing the speed of workers at gas stations. The problem with these businesses, I learned, is that they are not equivalent across nations. Gas stations cater to a very different clientele, and attract a different class of workers, in places like Indonesia and Brazil than in countries like the United States and Japan. Even in developed nations, service stations may be difficult to compare. Upon returning to the United States from the Far East, journalist Edwin Reingold reflected, "At what used to be called a service station [in the U.S.], the attendant, who sits behind bulletproof glass, can do nothing to help a novice learn the new greasy, smelly routine of pumping his own gas. Memories flood back of the typical Tokyo station, where a horde of neat, well-mannered, and expert attendants take charge of the car, fill it up, wash it, and check the tires. Then they doff their hats, shout their thanks, and stop traffic so the customer can drive away."[1] Clearly, timing the speed of workers in gas stations would tell something both more and less about a culture than its norms of temporality.

Eventually, three measures of the pace of life were developed: (1) walking speed—the speed with which pedestrians in downtown areas walk a distance of 60 feet; (2) work speed—how quickly postal clerks complete a standard request to purchase a stamp; and (3) the accuracy of public clocks. (These experiments are more fully described in Chapter 1.) My students and I have made these observations in as many countries as we have been able to get to. In a few countries, I have conducted the experiments myself; more often, the data have been collected by interested students from my university who were either traveling to foreign countries or returning to their home cities or countries for the summer. In all, we have collected data in at least one large city in each of 31 nations around the world.

THE PACE OF LIFE IN 31 COUNTRIES

The numbers in the table represent each country's ranks on the three measures. Lower ranks indicate faster walking speeds, faster post office speeds and more accurate public clocks. The overall pace of life score was calculated by statistically combining the times for each country on each measure:[2]

Country	Overall Pace of Life	Walking Speeds	Postal Times	Clock Accuracy
Switzerland	1	3	2	1
Ireland	2	1	3	11
Germany	3	5	1	8
Japan	4	7	4	6
Italy	5	10	12	2
England	6	4	9	13
Sweden	7	13	5	7
Austria	8	23	8	3
Netherlands	9	2	14	25
Hong Kong	10	14	6	14
France	11	8	18	10
Poland	12	12	15	8

(*continued*)

Country	Overall Pace of Life	Walking Speeds	Postal Times	Clock Accuracy
Costa Rica	13	16	10	15
Taiwan	14	18	7	21
Singapore	15	25	11	4
U.S.A.	16	6	23	20
Canada	17	11	21	22
S. Korea	18	20	20	16
Hungary	19	19	19	18
Czech Republic	20	21	17	23
Greece	21	14	13	29
Kenya	22	9	30	24
China	23	24	25	12
Bulgaria	24	27	22	17
Romania	25	30	29	5
Jordan	26	28	27	19
Syria	27	29	28	27
El Salvador	28	22	16	31
Brazil	29	31	24	28
Indonesia	30	26	26	30
Mexico	31	17	31	26

THIRTY-ONE COUNTRIES COMPARED

Japan and Western European countries scored fastest overall. Eight of the nine fastest countries were from Western Europe,[3] with Japan the lone intruder on this monopoly.

Switzerland achieved the distinction of first place, based on across-the-board high rankings: its walking speed ranked third, postal times ranked second, and—in one hell of a splendid finding, I must say—clock accuracy ranked first; their bank clocks were off by an average of a grand total of 19 seconds. Ireland ranked second, clocking in with the fastest walking speed of the 31 countries. (When I showed this result to a surprised Swedish colleague he at first shook his head and questioned, "Little Dublin is the fastest?" After a moment he smiled and said, "Yes, of course. That bloody chill keeps them moving.") Germany finished just behind, in third place, overall.

Japan was a close fourth. The three countries scoring ahead did so by very narrow margins—a few seconds here or there and the Japanese would have been in first place. There is, in fact, considerable evidence that Japan may be the fastest country of all. On the postal measure, for example, the Japanese had to settle for fourth, but where else besides Japan would our experimenter encounter postal clerks who sometimes wrapped the stamp in a little package, or, without being asked or required, sometimes wrote out receipts? We tried to correct for these extra seconds in our final tallies, but can one really give due credit to postal clerks who operate at near capacity speed while providing luxury service? The clerks in Frankfurt may have scored a few seconds faster, but it is difficult to imagine consumers there leaving the post office feeling like they had just made a purchase at Tiffany's. Or how about compared to China, where several clerks laughed at an experimenter, whom they apparently thought was crazy because he communicated with a note? And India, where we had to abandon our experiments because most clerks didn't believe it was their responsibility to carry change?

Or New York City? In the main post office (the proud owner of zip code 10001), one clerk held my note over her head, and proceeded to announce, very slowly and very loudly, to the line behind me and to much of the rest of midtown Manhattan: "YOU . . . MEAN . . . TO . . . TELL . . . ME . . . THAT . . . YOU . . . WANT . . . ONE . . . LOUSY . . . STAMP . . . AND . . . YOU'RE . . . GIVING . . . ME . . . A . . . [speaking even more slowly and loudly now, her cadence beginning to sound like the score from *Bolero*] . . . FIVE . . . DOLLAR . . . BILL?" After a short pause, and a handful of double takes at both the note and at me, she cranked up the volume a few more decibels, announcing: "GOD, HOW I HATE THIS CITY." Not only was this my most embarrassing moment as a researcher, but her speech so frightened me that I forgot to time her progress. (New York and Budapest were the only cities where experimenters reported being insulted by clerks.)

Whether Japan or Switzerland deserves the gold medal for speed remains an arguable issue, but without question the most remarkable finding at the front end of the rankings was the consistently fast scores from Western Europe. Eight of the nine Western European countries tested (Switzerland, Ireland, Germany, Italy, England, Sweden, Austria and The Netherlands) were faster than every other country other than Japan. The only "trailer" from Western Europe was France, which allowed Hong Kong (hardly a slouch in the hard-work category itself) to come in a notch in front of it. And even this minor slippage may have been the quirk of a rare environmental event: the Parisian measures were taken during the height of one of the hottest summers the city had ever experienced.

Before the study began, some of my colleagues predicted that one or more of the rapidly developing Asian economic powers would score the fastest. Michael Bond, an acclaimed cross-cultural psychologist from the Chinese University in Hong Kong, argued that his home culture would beat the field hands down. "The pace of life here [Hong Kong]," he proclaimed to a *Time* magazine reporter, "is a lot faster than anyplace else in the world."[4] With the assistance of Bond and his students we were able to gather several sets of reliable data in Hong Kong. But alas, Hong Kong not only scored behind Japan on all three measures, but was passed by virtually all of Western Europe. Hong Kong was, however, a bit faster than the three other industrialized Asian nations—Taiwan, Singapore, and South Korea—which scored fourteenth, fifteenth, and eighteenth respectively.

The United States, represented by its classic speedster New York City, was an unexpectedly slow sixteenth in overall pace. In fact, we were so surprised by the relative slowness of New York's scores that, as a reliability check, we sent out a second experimenter to collect a new set of observations; these turned out to be virtually identical to the first ones. New Yorkers did score a very respectable sixth place on walking speed, but were twenty-third on postal times and twentieth on clock accuracy.

Of course, straight-ahead speed may not be the single appropriate criterion for gauging the tempo of New Yorkers. One encounters a certain skill and assertiveness on the streets of New York that doesn't necessarily show up on a stopwatch. Whereas pedestrians in Tokyo are generally disciplined, meticulous, punctilious and even docile, New Yorkers are a study in anarchy. Sociologist William Whyte, who spent much of his career observing pedestrian behavior in large cities, pointed out the intensity with which New Yorkers challenge each other, disrespect red lights, are "incorrigible jaywalkers," zigzag between cars and bully vehicles, as if to say, "Make way for me or kill me." (Unfortunately, this bravado is often matched by that of the drivers. In 1994, for example, 12,730 New York pedestrians were hit by cars—approximately one accident every 41 minutes. Two hundred and forty-nine of these pedestrians were killed.[5]) So what if they fall a few tenths of a second behind Tokyoites? Loyalists might argue that the real speed of New Yorkers resides in the nuances. Whyte, for one, concluded: "It may be parochial of me, but I think that of all the pedestrians, New York's are the best."[6] Perhaps our rankings do, in fact, underestimate the pace of life experienced by people in large U.S. cities. I will add fuel to this contention shortly.[7]

WHERE LIFE IS SLOW

Down in Brazil
It takes a day to walk a mile,
Time just stands still.

MICHAEL FRANKS, *Sleeping Gypsy*

There were few surprises on the slow end of the list. The last eight ranks were all occupied by nonindustrialized countries from

Africa, Asia, the Middle East, and Latin America. The slowest of all were the great cradles of *amanhã*, rubber time, and *a mañana*: Brazil, followed by Indonesia, and, in last place, Mexico.

Slowness in these countries seeps into the very fabric of daily life. In our time surveys in Brazil, my colleagues and I found that Brazilians not only expected a casual approach to time, but had abandoned any semblance of fidelity to the clock. When asked how long they would wait for a late arriver to show up at a nephew's birthday party, for example, Brazilians said they would hold on for an average of 129 minutes. Over two hours! Among my own circle of parent friends, birthday parties are often planned to endure for a total of two hours. Forget about missing the beginning of the party. Parents who arrive 129 minutes after the scheduled start-up time are 9 minutes late for taking their children back home—a rather serious bit of temporal negligence to the host parent. And they had the same relaxed approach toward early arrivers: Brazilians allowed an average of about 44 minutes before dubbing someone early for this same birthday party, while respondents in the United States drew the line at 26.

For a lunch appointment, Brazilians said they would wait for an average of 62 minutes. Compare this to the United States, where people rarely allot more than one hour for lunch. On a workday, at least, typical Americans would need to be back at their office two minutes *before* the tardy Brazilian lunch was just getting ready to begin.

In Brazil, in fact, the midday meal is a very leisurely affair. For example, during my residence there I often took lunch at the home of a family of a Brazilian friend who at the time was the vice president of my university. Sometime between noon and 12:30 P.M. on those occasions my friend, decked out in a formal suit befitting his VIP role, would rush through the door, head for his bedroom, and quickly emerge in shorts and a t-shirt, suddenly relaxed and smiling—without fail, looking relaxed and smiling. We would all share a drawn-out feast, drink some wine or beer, chat a bit, play with the children or watch television until yawns began to appear. Each of us then retired to separate rooms for a substantial nap. Around 3:00 o'clock—*mais ou menos*, of course—my friend emerged from his room, back in his vice president's outfit, grimacing like the big shot that he was, and went off to continue running the university for a few more hours. Such is the luxury of the Brazilian *siesta*.

Fewer Brazilians wear watches than do people in the United States, and the watches they do wear are much less accurate. The predominance of inaccurate or nonexistent timepieces has also been incorporated into the culture of slowness. My favorite excuse from Brazilians who arrived late was, "*O relógio causou o meu atraso* [The clock caused me to be late]"—meaning that their delay was caused by a watch that was slow or set incorrectly.

Even a person with a first-rate watch in Brazil finds it difficult to be on time. Few people have their own cars, and public transportation is unreliable, to say the least. On more than one occasion, my bus driver abandoned his vehicle in the middle of our route. Once he returned after more than ten minutes, taking the last bite on a sandwich, and thanked us all for our patience. Another driver once excused himself for "*um momento*" and returned some fifteen minutes later with his groceries. In both cases I seemed to be the only person who was losing confidence. "*Calma, Bobby*" my companions would say to me, as they did in many situations during my Brazilian experience. Under these conditions, everyone is forced to take their time or go crazy, so that slowness becomes a self-fulfilling prophecy.

Many Brazilians we tested were completely out of touch with clock time. One watchless gentleman whom I asked the time looked me in the eye and proudly—a bit condescendingly, in fact—announced that it was "exactly 2:14." He was off by more than three hours. Compare this to one of my students in California who, when I asked him the same question, peeked at his watch and answered "Three-twelve and eighteen seconds."

In last-place Mexico, people who attend too closely to the clock can be a real nuisance. My colleague Sergio Aguilar-Gaxiola, with both an M.D. degree and a Ph.D. in clinical psychology, grew up in Mexico but has divided his professional life between Mexico and the United States "If you're invited to a party for a certain hour," Aguilar-Gaxiola observes, "it's understood that you should arrive late. If you show up at the scheduled hour—*en punto*—you may find yourself in the way of your hosts setting up or getting dressed. These rules about punctuality play a very important role in Mexican culture."

Slowness is so ingrained in Mexican culture that people who abide by the clock invite insult. "In Mexico," Aguilar says, "it's expected that people are going to be late, no matter what. If a meeting or gathering is scheduled for 11, *hora mexicana* is understood to be 11:15, 11:30, or even 12 o'clock, depending on the circumstances. If you arrive at 11, you not only expect to be by yourself, but to feel a certain embarrassment at being on time. Late arrivers—or punctual arrivers, depending on whether you are measuring by *hora mexicana* or *hora inglesa*—often tease the person who shows up right at eleven. There is a term, '*Llegaste a barrer*?', roughly meaning 'Did you arrive with the clean-up crew?', which is sarcastically spoken to the early arriver. It's a sharp putdown. In the United States, in fact, I still have difficulty mentally rescheduling my appointments from *hora mexicana* to *hora inglesa*—to arrive *en punto*—because I'm afraid of walking into that condescending '*Llegaste a barrer*?' taunt."

Some of what doesn't appear in the data tells even more about the slowness of the laggers on the list. On my first visit to the large central post office in next-to-slowest Jakarta, Indonesia, for example, I asked for the line to buy stamps and was directed to a group of private vendors sitting outside. Each of them hustled for my business: "Hey, good stamps, mister!" "Best stamps here!" In the smaller city of Solo, I found a volleyball game in progress when I arrived at the main post office on Friday afternoon. Business hours, I was told, were over. When I came back the next Monday, the clerk was more interested in discussing relatives in America. Would I like to meet his uncle in Cincinnati? Which did I like better: California or the United States? Five people behind me in line waited patiently. Instead of complaining, to my surprise, they began paying attention to our conversation. So not only did we find that the Indonesians move slowly but—between our needing to run extra trials because of experimental irregularities, along with having to get around by motorcycle and bicycle taxi—it took us considerably longer than in many other countries to find this out.

"Ah, where have they gone, the amblers of yesteryear?" asks Milan Kundera in his novel *Slowness*. Many of them, it seems safe to say, are on the streets of Jakarta, Rio de Janeiro, and Mexico City.

WHITHER LA DOLCE VITA?: WESTERN EUROPE, JAPAN, AND THE U.S.A. COMPARED

> "Will you walk a little faster?" said the whiting to a snail. "There's a porpoise close behind us, and he's treading on my tail."
>
> LEWIS CARROLL, *Alice in Wonderland*

A few years ago, *New York Times* journalist Alan Riding contrasted the compulsive workaholism of the United States and Japan to the ease with which much of Europe relaxes in the pleasures of the good life. Under the headline, "Why *La Dolce Vita* Is Easy for Europeans . . . As Japanese Work Even Harder to Relax," Riding asked: "How is it that Europeans sit around all day drinking coffee, spend long evenings over dinner, dress elegantly, get up late, take long vacations . . . Why, in short, do Europeans live so much better than Americans?"[8]

How do we reconcile the results of our experiments with this popular stereotype? Should we conclude from our data that *La Dolce Vita* of Western Europe is a dream of the past—that the Japanese and Western Europeans are the new stressed-out, time-urgent Type A's of the world while the United States has finally learned to relax? To answer this question, it may be helpful to look beyond our three measures of speed, which were designed to focus on facets of the tempo of workday life. What about the duration of this tempo? How long are people's off-hours? Do they enjoy vacations? What is the balance between hard work and leisure? It is here that Western Europe continues to diverge sharply from the United States, and even more from Japan.

To begin with, the average work week is shorter in most European countries than it is in the United States; both have shorter hours than Japan. One recent estimate indicates that the average annual paid working hours are 2,159 in Japan, compared to 1,957 in the United States, 1,646 in France, and 1,638 in the former West Germany. Workers in Japan, in other words, put in an annual average of 202 hours more than their counterparts in the United States and 511 hours more than workers in West Germany. Taking a 40-hour week as a base, these figures mean that the average Japanese salaryman spends five more weeks on the job than his colleagues in the United States and over twelve and one-half weeks—over three months!—more than workers from France and West Germany.[9] Looked at another way, only 27 percent of the Japanese labor force works as little as a five-day, 40-hour per week job, compared to 85.1 percent in the United States and 91.7 percent in France.

Italy, the font of *La Dolce Vita*, is a good example. The Roman postal workers were a respectable twelfth on our measure, which surprised me. After all, this is a country whose postal system has a reputation of being so bad that, not too long ago, there were publicized reports of trucks from Rome dumping loads of old mail into empty fields. On closer examination, however, there is evidence that the scores of the Romans on our test may not necessarily herald the arrival of a cutting-edge work ethic. A recent report by the government's authoritative Censis Foundation highlighted continued widespread waiting and delay in institutions like the post office. Our data only indicate that transaction times in Italy are speeding up behind the window; they say nothing about how long it takes to get to that window. Perhaps this combination of speedy times and long waits makes sense when one considers that post offices in Italy are only open for about five hours a day.[10]

It is notable that the difference in working hours between Western Europe and other first-world countries is widening. Until the

1940's the average hours in both Europe and the United States had been declining in tandem for nearly a century. In the United States, as in Europe, the issue of shorter hours was at the heart of the labor movement from the beginning; the question of work hours was once the "cause of the awakening" of the American laborer. "Eight hours for work, eight hours for sleep, eight hours for what we will," was the cry of turn-of-the-century unionists. Many of the most dramatic and significant events in the labor movement's history—for example, the strikes of 1886, the Haymarket riots, and the steel strike of 1919—were about the length of the work day. At first, even employers supported shorter hours—not out of any particular idealism, but because they were convinced that overwork and fatigue were counterproductive; that safety, health, rest, and a semblance of family life would pay for themselves over the long run. As a result, there was a gradual and steady decline in work hours in the United States throughout the latter nineteenth century, and a dramatic reduction—from ten-hour days to eight-hour days—during the first two decades of the twentieth century. Then the average work week was cut from six to five days, resulting in the forty-hour week.[11]

For a time it appeared that the downward trend in Americans' work time would continue. In 1930, for example, during the depths of the depression, economic visionary W. K. Kellogg (as in corn flakes), announced a revolutionary experiment: Nearly every employee in his huge Battle Creek plant would thereafter work a six-hour day. The reduction in hours was accompanied by only a minimal cut in pay, since Kellogg believed that hard work would replace long hours. Labor historian Benjamin Hunnicutt, in his book *Kellogg's Six Hour Day*, has documented that the program was an instant success. It was lauded by the media, business and labor leaders, and President Hoover himself. The reaction was typified by the front cover proclamation of one business magazine that this was "the biggest piece of industrial news since Ford announced his five-dollar-a-day policy."[12]

For nearly two decades, by nearly every yardstick, Kellogg's brainchild worked brilliantly. Workers appreciated the extra time.

Women, especially, reported that they enjoyed the added hours for activities like gardening, sewing, canning, caring for family members, and helping out in the neighborhood. Kellogg was equally pleased with the results. He reported that, as a result of the short schedule, overhead "cost was reduced 25% . . . labor unit costs reduced 10% . . . accidents reduced 41% . . . (days lost per accident) improved 51% . . . [and] 39% more people [were] working at Kellogg's than in 1929." Kellogg concluded that "with the shorter working day, the efficiency and morale of our employees is so increased, the accident and insurance rates are so improved, and the unit cost of production is so lowered that we can afford to pay as much for six hours as we formerly paid for eight."[13]

But in an increasingly work-obsessed nation, Kellogg's idyllic experiment was eventually doomed. In the aftermath of World War II, management came to subscribe to the view that, as one former Kellogg worker had quietly observed, "only an idiot would think you can get as much working less instead of more hours a week." Following the war, the company promoted a new policy that linked higher wages to greater productivity. Workers, hoping to cash in on the nation's post-war consumer bonanza, began to demand eight hours of work. Even the union fought for a return to eight hours. In a profound reflection of the national mood, Kellogg workers, management, and the union began to trivialize the notion of leisure; time off work was "wasted," "lost," "silly." Working for shorter hours became feminized. "Six hours was for the women," recalled one worker.[14] Those who held out for the old six-hour standard were "sissies," "lazy," or just "weird." Throughout the 1950s and 1960s, Kellogg employees steadily moved to an eight hour workday. In 1985, the few remaining holdovers, more than three-quarters of whom were women, surrendered. The milestone was barely reported in the media.

In the United States as a whole, the average work week has remained unchanged for more than half a century. In fact, many experts believe that leisure time has actually been decreasing. Historian Juliet Schor, for example, in her widely publicized book *The Overworked American: The Unexpected Decline of Leisure*, argues persuasively that the average American has less time to themselves than twenty years ago. This loss of leisure is not an accident. Schor presents evidence that unions in the United States have focused little attention on the question of working hours, preferring instead to direct their energies toward issues of salary and job security.

In Europe, on the other hand, the downward trend in work hours has hardly missed a beat. Unlike the United States, organized labor in Europe has kept the issue of shorter working hours at the top of its agenda throughout the postwar period. When economic crises hit, workers have fought the pressure for longer hours. In Germany, for example, a series of bitter strikes in the 1980's have earned a contract for a 35-hour work week for members of the large German union IG Metall. This standard is expected to spread throughout the German labor force. And so, after nearly one hundred years of simultaneous decline, the U.S. work week has remained flat, or perhaps even increased, over the last half century, while in Europe it persists in its sweet decline.

Workers in France—where work is sometimes viewed as an irritating, if necessary, interruption to living—are fighting for even more lenient contracts. In 1996, after French truck drivers snarled the country with a series of bitter strikes, the government conceded to lower their retirement age to 55. (For people like dancers and musicians in opera companies, the retirement age in France is now as low as 45 years). With that issue settled, unions are now focusing their attention on the length of the work week. As I write this sentence in January, 1997, six public transport unions have called a new strike to demand a 32-hour work week, with no loss in pay.[15] And so, after nearly one hundred years of simultaneous decline, the U.S. work week has remained flat, or perhaps even increased, over the last half century, while in Europe it persists in its sweet decline.[16]

Western Europe also leads the United States, and Japan, by an even wider margin in vacation time. In France, for example, workers by law receive at least five weeks and often six weeks of paid vacation. Every country in Europe, in fact, has collective bargaining agreements guaranteeing minimum paid vacations ranging from four to five and one-half weeks. In most cases, these mandated vacation periods range up to six weeks. In Sweden, it goes as high as eight weeks. Generous leave time is also provided for other purposes. In France, for example, it is official national policy to allow women 22 weeks of paid maternity leave and an additional year of unpaid leave.[17] The social welfare states of Scandinavia, where enhancing the psychological quality of life has long been a focus for both officials and the populace, go even further. In Sweden, for example, new parents are entitled to a combined twelve months leave of absence at nearly full pay, and another three months at reduced pay. Swedish parents are also entitled to 60 days per year (120 days in some cases), at 80 percent of their normal pay, to care for a sick child.[18]

In the United States, on the other hand, vacation time for most workers remains limited to the traditional two weeks—that is, if they have been fortunate enough to avoid being shifted to seasonal contracts, in which case they may get no paid vacation at all. Harris polls indicate that Americans report a 37 percent decrease in their leisure time over the past twenty years[19].

Even the time for bereavement in America has shrunk in the last century. In 1927, Emily Post prescribed the appropriate formal grieving period for a widow to be three years. As of 1950, that period had diminished to three months. By 1972, Amy Vanderbilt advised that the bereaved "pursue, or try to pursue, a usual social course within a week or so after a funeral."[20] Sociologist Lois Pratt, in a study of the bereavement policies of American business enterprises, found that the majority of companies now limit official bereavement leaves to about 72 hours. Employees are expected to complete their mourning during these three days and then return to business as usual.[21] The guidelines for appropriate bereavement periods sometimes get very specific. In the case of weekend deaths, for example, the American Management Association has drafted the following guidelines: "It is usually expected that when death occurs on a Saturday the employee should return to work on a Tuesday following the normal time for the funeral."[22]

In Japan, vacation time is even scarcer than it is in the United States. Although the average number of paid vacation days offered in Japan hovers around a respectable three weeks, the Japanese Ministry of Labor reports that only about half of this time is actually used. In 1990, for example, an average of 15.5 days of vacation time were authorized, of which 8.2 days were taken.[23]

Statistics from a 1989 Eurobarometer survey confirm the impression that Europeans are more comfortable with the time in their lives. As part of this survey, respondents from each of the then 12 European Union countries were asked how they felt about "the time you have available to do things that have to be done." Averaging across the 12 countries, 83.4 percent of all respondents reported that they felt "very good" or "fairly good" about their available time.[24] Whither *La Dolce Vita?* Apparently, where it began—in Western Europe, where workers have not only mastered the art of speed and productivity at work, but seem to have managed to use it to retain at least some remnant of the good life in their leisure hours.

Responses to similar survey questions in Japan and the United States portray a very different image. In the United States, a 1992 survey by the National Recreation and Park Association found that more than one-third of all Americans (38 percent) complained that they "always felt rushed." This percentage of people who feel chronically rushed is sharply up from 22 percent in 1971 and up slightly from 32 percent in 1985, as reported in surveys conducted by John Robinson's Americans' Use of Time Project.[25] Other surveys emphasize the extent to which most U.S. workers crave free time. In a 1991 national survey of time values conducted by the Hilton Hotel Corporation, two-thirds of all respondents said they would accept salary cuts if it meant getting more time off from work. This two-thirds figure was relatively consistent across gender, age, educational background, economic status, and number of children. The most recent survey data indicate that Americans' sense of feeling pressed for time may finally be stabilizing.[26] Still, it is clear that workers in the United States feel they put in more than their share of hours; on the other hand, the tempo data from our experiments raise the question of how much is accomplished during these hours.[27]

In Japan, people offer an even dimmer appraisal of their available time. When a 1991 survey conducted by the *Mainichi Shimbun* newspaper asked people "To what extent do you feel pressed for time in your daily life?" a total of 70 percent of all respondents said they felt "somewhat" (44 percent) or "strongly" (26 percent) time pressured. These figures were almost identical for men and women.[28] In a 1989 survey conducted by the *Yomiuri Shimbun* newspaper, only 12 percent of all Japanese said that they were at least pretty satisfied with their leisure time. Then again, in the same survey, only 24 percent said that leisure time played a particularly important role in their overall life satisfaction, an issue I will return to later.[29]

THE PACE OF LIFE IN 36 U.S. CITIES

The most vivid examples of cultural differences tend to be seen when comparing countries. But cities and regions within countries can also differ enormously. This is certainly true for the United States. On the surface, areas of the United States may appear relatively homogeneous; most people speak the same language and almost everyone shares most of the major systems of communication. But the native of New York City who is transferred to the Deep South, or the Midwesterner who moves to Los Angeles, had best be prepared for a high-voltage culture shock.

Even short geographical shifts can be profound. When the writer Jan Morris moved from Oklahoma to Texas, it was, she felt,

> like entering France, say, out of Switzerland. The moment I crossed the Red River out of Oklahoma the nationality of Texas assaulted me, almost xenophobically, and I seemed to be passing into another sensibility, another historical experience, another set of values, perhaps.[30]

Are there significant differences in the pace of life between the cities of the United States? Do New Yorkers live up to their reputation of living in the fast lane? Are Californians really more laid-back than people in other parts of the country? Which are the fastest and the slowest places?

To answer these questions, my students and I collected four measures in each of 36 cities across the United States.[31] We selected these particular cities to target a representative cross-sample of the nation's metropolitan areas. From each of the four census-defined regions of the United States—the Northeast, Midwest, South, and West—we studied three large metropolitan areas (population greater than 1,800,000), three medium-sized cities (population between 850,000 and 1,300,000) and three smaller cities (population between 350,000 and 560,000).[32]

Our measures of pace of life were variations on those in the 31-country study. First, we again clocked the walking speeds of pedestrians, walking alone, over 60-foot distances in downtown locations during business hours. Second, as an indicator of work speed, we timed how long it took bank clerks to either give change, in set denominations, for two $20.00 bills, or to give us two $20.00 bills for change. (Requests for making change were alternated with those for giving change simply so that experimenters would not be running around with wads of money in their pockets.)[33] Third, as a measure of people's concern with clock time, we counted the proportion of randomly selected men and women in downtown areas who were wearing wristwatches during business hours. In most cases subjects were wearing short sleeves, allowing us to easily observe whether they were wearing watches. People in long sleeves were approached and asked the time of day. We also added a fourth indicator of the pace of life—talking speed. In each city we tape-recorded the responses of postal clerks to a standard question—to explain the difference between regular mail, certified mail, and insured mail. Later, research assistants played back the tapes and calculated "articulation rates" by dividing the number of uttered syllables by the total time of the response.

FAST AND SLOW CITIES IN THE U.S.A.

Lower ranks indicate faster speeds and more watches worn.

	Overall Pace [34]	Walking Speed	Bank Speed	Talking Speed	Watches Worn
Boston, MA	1	2	6	6	2
Buffalo, NY	2	5	7	15	4
New York, NY	3	11	11	28	1
Salt Lake City, UT	4	4	16	12	11
Columbus, OH	5	22	17	1	19
Worcester, MA	6	9	22	6	6
Providence, RI	7	7	9	9	19
Springfield, MA	8	1	15	20	22
Rochester, NY	9	20	2	26	7
Kansas City, MO	10	6	3	15	32
St. Louis, MO	11	15	20	9	15
Houston, TX	12	10	8	21	19
Paterson, NJ	13	17	4	11	31
Bakersfield, CA	14	28	13	5	17
Atlanta, GA	15	3	27	2	36
Detroit, MI	16	21	12	34	2
Youngstown, OH	17	13	18	3	30
Indianapolis, IN	18	18	23	8	24
Chicago, IL	19	12	31	3	27
Philadelphia, PA	20	30	5	22	11
Louisville, KY	21	16	21	29	15
Canton, OH	22	23	14	26	15
Knoxville, TN	23	25	24	30	11
San Francisco, CA	24	19	35	26	5
Chattanooga, TN	25	35	1	32	24
Dallas, TX	26	26	28	15	28
Oxnard, CA	27	30	30	23	7
Nashville, TN	28	8	26	24	33
San Diego, CA	29	27	34	18	9
East Lansing, MI	30	14	33	12	34

	Overall Pace	Walking Speed	Bank Speed	Talking Speed	Watches Worn
Fresno, CA	31	36	25	17	19
Memphis, TN	32	34	10	19	34
San Jose, CA	33	29	29	30	22
Shreveport, LA	34	32	19	33	28
Sacramento, CA	35	33	32	36	26
Los Angeles, CA	36	24	36	35	13

THE SPEEDY NORTHEAST

In general, our results confirm the widespread impression that the Northeastern United States is fast-paced, whereas the West (more accurately, California, which accounted for eight of the nine western cities) is more relaxed. The three fastest cities, and seven of the nine fastest, were from the Northeast. Northeasterners generally walk faster, give change faster, talk faster, and are more likely to wear watches than people in other U.S. cities.

Boston edged out Buffalo for first place. New York City, the pre-study favorite, was a close third. But perhaps New Yorkers may be allowed a few tenths of a second to make up for the local festivities that occurred during data collection. Walter Murphy, who gathered the walking speed data there, reported that he had to close up shop at one point because of an improvised music concert that materialized while he was timing walkers. Then, after moving to a new location, Murphy encountered an attempted purse snatching followed by an unsuccessful mugging. All of this occurred during a period of one and a half hours. New York pedestrians maneuvered through all of this with their trademark assertiveness, demonstrating circuslike skills not counted in the final tallies.

The California cities had the slowest pace overall, due mostly to particularly slow walkers and bank tellers. The very slowest times belonged to America's symbol of sun, fun, and laid-back living: Los Angeles. Los Angelenos were twenty-fourth in walking speed, next-to-last in speech rate, and far, far behind every other city we studied in bank teller speeds. Their only concession to the clock, in fact, was to wear one—the city was thirteenth highest in watches worn. Californians' relaxed attitudes toward time also show in other ways. To obtain the exact time of day in most California cities, for example, you dial the phone number spelling out "P-O-P-C-O-R-N." Compare this to first-place Boston, where the number for the same information is "N-E-R-V-O-U-S."

The biggest problem for our experimenters in many of the West Coast cities—Los Angeles and my home of Fresno being prime examples—was the walking speed experiment, where we were hard pressed to find any walkers at all. Most of the pedestrian traffic in these suburban cities seemed to be limited to parking lots and shopping malls, neither of which are comparable to the main business centers we studied in other cities. And often the people in downtown areas who did traverse a full 60 feet were (I certainly hope) less than representative of the suburban populations as a whole: they were the homeless, the unemployed, and mobile hookers. We saw joggers, bicyclists, rollerbladers, and even a few old-fashioned skateboarders, but most of the pedestrians we observed went no further than their parked cars. The true public pathway in most of these places is the highway. Joan Didion may not have been far off when she observed that the freeway experience is Los Angeles's only secular form of communion. After considerable time, we did eventually manage to locate minimally sufficient samples of pedestrians in downtown locales, but I must admit that at several points I was tempted to just go down to the local health club to clock people on their treadmills.

How much of a difference was there between cities? Often not a great deal from one rank to the next. At the extremes, though, people march to very different drummers. In walking speed, for example, the fastest pedestrians—in Springfield (11.1 seconds) and Boston (11.3 seconds)—covered the 60-foot distance an average of 3.5 seconds faster than those in Chattanooga (14.6 seconds) and Fresno (14.7 seconds). In other words, if they were walking a football field, the teams from Massachusetts would be crossing the goal line at about the same time their Californian opponents were still about 25 yards short.

Differences in talking speed were even greater. The fastest-talking postal clerks—in Columbus, Ohio (3.9 syllables/second)—uttered nearly 40 percent more syllables per second than their colleagues in Sacramento (2.9 syllables/second) and Los Angeles (2.8 syllables/second). If they had been reading the 6 o'clock news, it would have taken the Californian workers until nearly 7:25 to report what the Ohio workers concluded at 7:00.

Broken down by regions, the fastest overall times were in the northeast, followed by the midwest, the south and then the west. This fit predictions. Certainly, the difference between the East and West coasts confirms a popular stereotype. When Horace Greeley advised young men to go west, he was thinking of adventure and economic opportunity. But most of the Easterners I know who have migrated there were searching for a less frantic, more manageable style of life. I enjoy watching the reactions of New Yorkers who visit me in Fresno. Some are impressed by the wonderful life I've discovered, while others think I should see a neurologist. Typically, the approvers find my adopted home relaxing; they often comment on how many more hours there seem to be in the day in Fresno. The malcontents spend a lot of time asking what people do for fun in this town. (I'll have something to say about the issue of fitting people to environments later.) In both cases, the slower pace of life in Fresno is not only obvious to my guests, but the very locus of differences between our hometowns.

THE HONKOSECOND

Of course, there are many ways to evaluate the pace of life. Alternative experiments might lead to different results, as my critics have not been shy about pointing out. One of my favorite suggestions came from a Los Angeles journalist:

How about measuring how many drivers make sudden left turns in front of you? Or how quickly a police officer flags you down for a jaywalking ticket? Or how fast an automatic bank teller machine displays the "closed" sign just as you're about to insert your ATM card? Or the percentage of men and women wearing beepers? . . . Under our set of indicators, we're willing to bet that Los Angeles would rank the fastest-paced city in the country.[35]

A UCLA professor suggested I should measure "the honkosecond . . . the time between when the traffic light changes and the person behind you in L.A. honks his horn." The honkosecond, he claimed, is "the smallest measure of time known to science."

Each of our experimental measures have their quirks. The number of people wearing a watch, for example, reflects not only a society's preoccupation with time but also its sense of fashion[36] and perhaps its level of affluence. Basing measurements on interactions with postal clerks and bank tellers puts undue emphasis on these rather specialized subpopulations; and the performance of the clerks and tellers depends on their skill and knowledge as well as on their general tendency to hurry or dawdle. Taken together, though, they sample a wide range of people and activities, and reflect many facets of a city's pace of life.

SEVEN

HEALTH, WEALTH, HAPPINESS, AND CHARITY

Beyond doubt, the most salient characteristic of life in this latter portion of the nineteenth century is its SPEED,—what we may call its hurry, the rate at which we move, the high-pressure at which we work;—and the question to be considered is, first, whether this rapid rate is in itself a good; and, next, whether it is worth the price we pay for it—a price reckoned up, and not very easy thoroughly to ascertain.

W. R. GREG, *Life at High Pressure* (1877)

I am forever fascinated by the fact that some *places* are faster than others. But if you are—as I am—the sort of person who makes a pastime of searching for greener pastures, the truly important issues go beyond speed ratings to questions about where people are better off. Where are they healthier? Happier? More charitable?

When answering these questions, it is tempting to assume that slow is healthy and fast is unhealthy—that the quality of life in places where people work quickly and work a great deal is inferior to that among their more leisurely counterparts. The image of frenetic overachievers working themselves to death in one setting, contrasted with dancing Zorba-the-Greeks happily embracing every moment of their time in others, makes for a tidy stereotype. But cultural values, especially ones as profound as those about time, rarely separate into such orderly categories of good and bad.

The pace of life does, in fact, have vital consequences for the quality of life. It casts its shadow on the physical and psychological health of individuals and upon the social well-being of communities. But the consequences are most often a "good news/bad news" story. Any given pace carries mixed blessings.

PHYSICAL WELL-BEING: THE TYPE-A CITY

Those who rush arrive first at the grave.
Spanish proverb

In the mid-1950's, two San Francisco cardiac specialists, Meyer Friedman and Ray Rosenman, noticed that the heart patients in their waiting rooms seemed more tense than other patients. More precisely, Friedman and Rosenman credit their insight to an upholsterer who called to their attention the peculiar fact that the chairs in their waiting room were only worn on the front edge of the seat. Acting on a hunch, they began a research program to explore the almost unexplored possibility that psychological stress might significantly contribute to the likelihood of a heart attack. At the time, the predominant belief in the medical community was that treating coronary artery disease was purely a mechanical matter; as one heart surgeon said, "It is a matter of plumbing."

In one early study, Friedman and Rosenman measured the blood cholesterol level of accountants from January through June. The accountants' eating and exercise habits did not change during the period. Yet during the first two weeks of April, as the stress of the April 15 income tax deadline approached, their average blood cholesterol level rose abruptly, and their blood clotting rate increased. In May and June, these measures had returned to normal.

Friedman and Rosenman reasoned that some people live in a self-imposed mindset of chronic tension. Psychologically speaking, the stressed-out patients in their waiting rooms are always in an accountant's mid-April. Subsequent research—most notably the Western Collaborative Group Study, which monitored the health and sickness patterns of 3,500 men over an eight and one half year period—has confirmed this assertion. More specifically, it has been found that patients with coronary heart disease are prone to a behavior syndrome characterized by a sense of time urgency, hostility, and competitiveness. People with this "Type A" behavior pattern are seven times more likely than "Type B's" (normals) to show evidence of heart disease and are more than twice as likely to have a heart attack.[1]

A hurried pace of life is a defining element of the Type A pattern.[2] Type A's tend to walk and eat quickly; to take pride in always being on time; and to juggle several activities simultaneously. They have little patience for others' "slow" behavior; for example, Type A's have a habit of completing sentences for speakers who take too long to come to the point. And, of course, Type A's work longer hours than Type B's.

With these Type A studies in mind, it seemed logical to question whether the pace of life in different places might be related to the prevalence of coronary heart disease in their populations. Is there such a thing as a Type A city? To answer this question, we looked at the pace-of-life scores from our experiments for different countries and cities, and compared these scores to the rates of death from coronary heart disease in these places.[3]

Our data strongly support the notion that cities, too, can be Type A. Faster places were much more likely to have higher rates of death from heart disease. This was true in our study of 31 different countries as well as of 36 cities in the United States. Not only did our results show a clear relationship between pace of life and heart disease, but the magnitude of the relationship was even higher than that usually found between heart disease and Type A personality tests at the individual level. In other words, our data suggest that the speed of a person's environment is at least as good a predictor of whether he will die of a heart attack as his scores on a Type A personality test.[4]

Why are people in fast places more likely to die of coronary heart disease? It is certainly unlikely that simply walking fast or working hard causes heart attacks. If that were the case, the streets would be littered with gasping jaywalkers and fallen joggers. In the United States, certainly, where people pay good money to work out on treadmills for the express purpose of keeping their circulatory system in good order, it seems contradictory to equate strenuous physical exercise with physical disease.

The main reason that heart disease is common in fast cities, I suspect, is that these places both attract and create an inordinate concentration of Type A individuals. Fast places appeal to fast people, and fast people create fast places. The social psychologist Timothy Smith and his colleagues have shown that Type A's both *seek* and *create* time-urgent environments. The fastest cities in our study may be both their dearest dreams and most ingenious creations.[5]

So the development of a fast-paced city could be explained by the following scenario. First, Type A's are attracted to fast-paced cities, resulting in more Type A residents. In turn, these Type A's do their best to drive the pace faster and faster. Meanwhile, slower Type B individuals tend to migrate away from fast-paced cities to more relaxed settings. Within fast-paced cities, Smith's research suggests, time urgency is expected of everyone—Type A's and B's alike. The result is that Type B's try to meet the demands of Type A's, and Type A's strive to push the beat even faster—all in an environment where migration patterns have already produced an overrepresentation of coronary-prone personalities.

Type A cities are stressful places. The pressures of time urgency may lead to unhealthy behaviors—the use of tobacco, alcohol, and drugs, an unhealthy diet, and a lack of healthy exercise. These, in turn, put people directly at risk for the development of coronary heart disease. Our studies as well as statistical information from the U.S. Department of Health and Human Services show, for example, there is more cigarette smoking in faster cities and countries. In the United States, we found that smoking rates follow the same regional patterns as our figures for coronary heart disease and the pace of life: smoking and coronary heart disease rates are highest, and the pace is fastest, in the Northeast, followed by the Midwest, the South and then the West.[6]

Anecdotal support for the role of smoking comes from a city that deviates sharply from the Type A city model. Salt Lake City, with its predominantly Mormon population, tested as the fourth highest American city in pace of life but is thirty-first (sixth from the lowest) in coronary heart disease. This may be largely explained by the fact that the Mormon religion strongly encourages hard work, but strictly prohibits cigarette smoking. Eric Hickey, a professor of criminology who is extremely active in the Mormon Church, explains that "Mormonism is a 24-hour, seven-day-a-week commitment. When you combine this with our family lives and our jobs, it doesn't surprise me that Salt Lake City has a very fast pace. But at the same time, Mormons are very spiritual people. We believe in moderation in most things; but for more physically debilitating activities—such as consuming alcohol, caffeine, smoking or drugs—we just don't do it at all." So in Salt Lake City, Mormon values act as a buffer to counteract a fast pace of life; the felicitous result is less coronary heart disease.

PSYCHOLOGICAL WELL-BEING: WHERE ARE PEOPLE HAPPIER?

It might stand to reason that a slow pace of life also makes for happier people. The issue conjures up the dreamy image of happy natives—presumably photographed by anxious but considerably more affluent tourists—in lazy villages next to timeless, exotic beaches. On the other hand, research has shown that economic productivity is highly related to people's happiness. This turns out to be true whether we are studying the economic well-being of individuals or countries: on the whole, wealthier people are happier and people in wealthier countries are happier. In one recent study, for example, personality psychologist Edward Diener and his colleagues at the University of Illinois found that the average life satisfaction in nations was highly correlated with a wide range of national economic indicators, including gross domestic product (GDP), purchasing power, and the fulfillment of basic needs.[7]

Since our own study found a very strong relationship between economic vitality and the pace of life, we hypothesized that this should also lead to a positive relationship between the pace of life and happiness. And this is exactly what we found: in all of our pace-of-life experiments, people in faster places were more likely to be satisfied with their lives.[8]

These results depict an apparent paradox: People in faster places are more prone to suffer coronary heart disease, but they are also more likely to be happier with their lives. If a fast pace of life creates the stress that leads to cigarette smoking and heart attacks, shouldn't this same stress make for an unhappy existence?

The root of this seeming inconsistency, I believe, is economics, and the cultural values that come along with it. Cultures that emphasize productivity and making money typically create a sense of time urgency and a value system that fosters individualistic thinking; and that time urgency and individualism in turn make for a productive economy. These forces—economic vitality, individualism, and time urgency—have both positive and negative consequences for people's well-being. On the one hand, they create the stressors that lead to unhealthy habits like cigarette smoking and coronary heart disease. On the other hand, they provide material comforts and a general standard of living that enhance the quality of life. Productivity and individualism—which in themselves are very difficult to separate from each other—have double-edged consequences.

As the economist Juliet Schor wrote in *The Overworked American*:

We have paid a price for prosperity. Capitalism has brought a dramatically increased standard of living, but at the cost of a much more demanding worklife. We are eating more, but we are burning up these calories at work. We have color televisions and compact disc players, but we need them to unwind after a stressful day at the office. We take vacations, but we work so hard throughout the year that they become indispensable to our sanity. The conventional wisdom that economic progress has given us more things *as well* as more leisure is difficult to sustain.[9]

It is not surprising to learn that these ostensibly paradoxical consequences are also found for other behaviors. Diener and his colleagues, for example, have observed that although divorce is much higher in individualistic nations, marital satisfaction is also often higher—the United States being a primary case in point. Their research has also found that suicide and psychological well-being are both higher in individualistic cultures than collective ones.[10]

It has been observed that the Chinese character for "crisis" is composed of the character for "danger" plus that for "opportunity." And the word "crisis" in our own language derives from the Greek word for "decision." In a similar vein, the fruits of individualism and hard work provide the potential for both psychological wealth and disaster.[11] Ultimately, how we structure our time is a choice between alternatives. A rapid pace of life is neither inherently better nor worse than a slow one.

SOCIAL WELL-BEING: WHERE DO PEOPLE HELP?

Not only does the pace of life affect psychological and physical well-being, but it may have important implications for the way people treat each other. Slowness is a social norm and, like other norms, may permeate deeply into accepted codes of conduct.

The Kelantese people of the Malay Peninsula, for example, put an emphasis on slowness that is deeply embedded in their beliefs about right and wrong. Haste is considered a breach of ethics. The

Kelantese are judged by a set of rules for proper behavior known as *budi bahasa*, or the "language of character." At the core of this ethical code is a willingness to take the time for social obligations, for visiting and paying respect to friends, relatives and neighbors. Any hint of rushing smacks of greed and too much concern for material possessions. Most important, it shows an irresponsible lack of attention to the social obligations of the *budi bahasa*. Violators threaten basic village values concerning interpersonal relations and village solidarity. They are gossiped about, considered less refined (*halus*), and are often suspected of trying to hide something.[12]

The sensibility of the *budi bahasa*—that rushing and making time for people are mutually exclusive—is widely appealing. It is certainly effective for the Kelantese. But the Kelantese are a nonindustrialized people; they have relatively few demands on their time. They also live in small villages where everyone is a neighbor, so that prosocial behavior is motivated by the knowledge that how one treats others will very likely be returned in kind. This raises the question of whether the *budi bahasa* philosophy holds for industrialized and more populous places, environments where most daily contacts are with strangers whom people will probably never meet again. Do the pace of life and social responsibility also go hand in hand in industrialized, urban settings?

Most contemporary urban theorists believe that they do. They argue that this connection between time and social behavior is not so much driven by moral questions, as it is for the Kelantese, but by social psychological reality. Social psychologist Stanley Milgram believed that the rapid pace of life in modern cities confronts people with more sensory inputs than they are able to process, creating what he calls psychological overload. The larger the city, the greater the overload. In order to adapt to this predicament, the overloaded urbanite screens out everything not essential to his goals. In essence, the city dweller focuses on his or her goals and moves directly toward them as quickly as possible. They have neither the time nor the psychological energy to attend to people who are peripheral to their lives. Strangers are especially prone to be ignored because of this screening process: for Milgram, the rapid pace of life of big cities virtually requires a disregard for the needs of strangers.[13]

Are people in fast cities less likely to take the time to help a stranger, as Milgram hypothesizes? Over the past several years, my students and I have conducted a series of studies to test this prediction. In the United States, my students and I returned to the same 36 cities[14] as in our pace-of-life research. This time we observed helping behavior, in a total of six situations:

Retrieving a dropped pen. The experimenter (a neatly dressed college-age male), walking at a moderate pace, would reach into his pocket and "accidentally," without appearing to notice, drop his pen behind him, and continue walking. In each city, we observed the number of occasions a passing pedestrian helped the experimenter retrieve the pen.

Hurt leg. Walking with a heavy limp and wearing a large and clearly visible leg brace (the ugliest ones we could find), the experimenter "accidentally" dropped, and then unsuccessfully struggled to reach down for, a pile of magazines. What proportion of approaching pedestrians offered assistance?

Blind person crossing the street. An experimenter wearing dark glasses and carrying a white cane acted the role of a blind person needing help getting across the street.[15] We measured the percentage of instances in which help was offered.

Change for a quarter. With a quarter in full view, the experimenter approached a pedestrian passing in the opposite direction and asked politely for change for a quarter. We observed how many pedestrians in each city stopped to check for change.

Lost Letter. A neat hand-written note, "I found this next to your car," was placed on a stamped envelope addressed to the experimenter's home. The envelope was then left on the windshield of a randomly selected car parked at a meter in a main shopping area. How many of these letters arrived at the address?

United Way Contributions. As a measure of charitable contributions, we looked at how much each city contributed, per capita, to United Way campaigns.[16]

In which U.S. cities are people more willing to take time to help a stranger in need? After statistically combining each city's scores on the six measures, we found the following ranks, from most to least helpful[17]:

1. Rochester, New York
2. Lansing, Michigan
3. Nashville, Tennessee
4. Memphis, Tennessee
5. Houston, Texas
6. Chattanooga, Tennessee
7. Knoxville, Tennessee
8. Canton, Ohio
9. Kansas City, Missouri
10. Indianapolis, Indiana
11. St. Louis, Missouri
12. Louisville, Kentucky
13. Columbus, Ohio
14. Detroit, Michigan
15. Santa Barbara, California
16. Dallas, Texas
17. Worcester, Massachusetts
18. Springfield, Massachusetts
19. San Diego, California
20. San Jose, California
21. Atlanta, Georgia
22. Bakersfield, California
23. Buffalo, New York
24. Salt Lake City, Utah

25. Boston, Massachusetts
26. Shreveport, Louisiana
27. Providence, Rhode Island
28. Philadelphia, Pennsylvania
29. Youngstown, Ohio
30. Chicago, Illinois
31. San Francisco, California
32. Sacramento, California
33. Fresno, California
34. Los Angeles, California
35. Paterson, New Jersey
36. New York, New York

Broken down by regions, helping tended be highest in the Southeast, followed by the Midwest, followed by the large cities in the Northeast—pretty much the inverse of the pace of life in these regions, which is what Milgram's system overload theory predicts. But there was one glaring exception. The 11 California cities, taken as a whole, were slowest on the pace-of-life experiments but the least generous with their time on the helping experiments.[18] And their per capita contributions to United Way were less than one-tenth of those from front-runner Rochester.

The state of New York is home to the most (Rochester) and least (New York City) helpful of the 36 cities. Harry Reis, a professor of psychology at the University of Rochester for more than 20 years who grew up in New York City, was "not the least bit surprised" by the performance of his two homes. "Rochester is a town where the social fabric hasn't deteriorated as much as in other places. People are rarely too busy to offer a helping hand."

But although cities like Rochester and New York fit Milgram's predictions, the unfriendly behavior we observed in most California cities underscores how complex a culture's norms of social responsibility really are. These Western cities demonstrate that having the time to lend a helping hand does not mean that that hand will be offered. Beliefs about the allocation of time and appropriate social behavior must be linked together by a moral code, as they are for the Kelantese, if we expect people to use their "extra" time for the welfare of others. Without such a code, a slow pace of life may lead to nothing more than being relaxed.

We see this when comparing the cities of the South to those in the West. Although both regions share an unhurried pace of life, very different sets of social values drive their temporal norms. In the case of the South, taking one's time tends to be connected with the notion of Southern ladies and gentlemen. It is tuned to a culture of gentility and civility. In the West, on the other hand, slowing down is mostly about—well, taking it easy. The laid-back norm often has less to do with loftier social values than with simply leading the good life.

Popular music captures this distinction nicely. When Hank Williams, Jr., sings, "We say grace, we say 'ma'am,'" everyone understands he's not talking about Southern California. And when the Beach Boys sang "She'll have fun, fun, fun 'til her Daddy takes the T-Bird away," where else could it be but Southern California? The stereotype was neatly summed up by one Nashville resident who commented about our results: "Here, we say, 'How are you?' In L.A. they say, 'How's your car?' And in New York they say, 'Give me your car.'"

In order for a slow pace of life to result in greater attention to the needs of strangers, it must be accompanied by a code of social responsibility—as the Kelantese have accomplished with their *budi bahasa*. A slow pace of life in itself may buffer against heart attacks, but it has no inherent moral philosophy. Not surprisingly, five of the seven most helpful cities were from the South, where there is such a code. In fact, if we reranked helpfulness with just the five face-to-face helping measures (excluding United Way contributions), the Southern cities would score even higher.

Lynnette Zelezny, now a psychology professor at my university, grew up in several locales in the South. "In the South," she observes, "it's the social norm to take the time to be kind to people— at least to be kind on the surface. Even if you aren't a churchgoer, that is still the social norm. When I was a child, if I wasn't acting appropriately in public—that's the key element, 'in public'—my grandmother would admonish me to 'be sweet,' which meant 'be kind.' We were expected to show what a kind family we were by our behavior. In the South, it's very important to wear this niceness on your shirtsleeve. Let them know that you're kind. If you run into an acquaintance at the grocery store here in California, you might say hello. In the South, you are expected to take the time to make light conversation not only with whoever else is in line but with the person who is checking you out."

Jean Ritter, also a colleague on my faculty, grew up in Little Rock. "The pace of life in the traditional South is slower," Ritter agrees. "It's related to etiquette. It's improper to rush or interrupt others or to neglect the social niceties no matter how time-consuming they might be. For example, when waiting in line, people are expected to be patient and to visit pleasantly with those around them. It is a gentler way of life in many respects. People take the time to stop and smile and greet each other on the street. And if you don't do that, you are considered rude. At the small college where I was an undergraduate student, it was expected that you spoke to every single person you passed on the sidewalk, and if you did not do that there was something wrong with you. There's a norm to take time to be friendly. It's completely different in California."

We also learned that there may be a difference between helping and civility. People in faster places were often less likely to act civilly even while they were helping. In New York City, helping often appeared with a particularly sharp edge. During the dropped pen experiment, for example, helpful New Yorkers would typically call back to the experimenter that he had dropped his pen, then quickly move on in the opposite direction. Helpers in the higher-scoring Southeastern cities, on the other hand, were more likely to return the pen personally, sometimes running to catch up with the experimenter. In the blind person situation, helpful New Yorkers would often wait until the light turned green, then tersely announce to the experimenter that it was safe to cross as they quickly walked ahead. In the Southeast, helpers were more likely to offer to walk the blind person across the street, and sometimes asked if he then needed further assistance.

In general, it often seemed as if New Yorkers were willing to offer help only when it could occur with the assurance of no further contact, as if to say "I'll meet my social obligation but, make no mistake, this is as far as we go together." How much of this is motivated by fear and how much by simply not wanting to waste time is hard to know. But in more helpful cities, like Rochester and much of the Midwest and South, it often seemed that human contact was the very motive for helping. People were more likely to help with a direct smile and to welcome the "thank you."

Perhaps the most vivid example of uncivil helping occurred with the lost letter measure. In many cities, I received envelopes that clearly had been opened. In almost all of these cases, the finders had then resealed the envelope or remailed it in a new envelope. Sometimes they had attached notes, usually apologizing for opening our letter. Only from New York did I receive an envelope which had its entire side ripped and left open. On the back of the letter the helper had scribbled, in Spanish: "*Hijo de puta irresposable*"—which, translated, makes a very nasty accusation about my mother. Below that was added a straightforward English-language "F___ You." It is interesting to picture this angry New Yorker, perhaps cursing my irresponsibility all the while he was walking to the mailbox; yet, for some reason, feeling compelled to take the time to perform his social duty, for a stranger he already hated. Ironically, of course, this rudely returned letter counted in the helping column for New York's score.

Compare this to a note I received on the back of a returned letter from Rochester:

Hi. I found this on my windshield where someone put it with a note saying they found it next to my car. I thought it was a parking ticket. I'm putting this in the mailbox 11/19. Tell whoever sent this to you it was found on the bridge near/across from the library and South Ave. Garage about 5 P.M. on 11/18.

P.S. Are you related to any Levines in New Jersey or Long Island?
L.L.

Nonhelpers also differed in their degree of civility. In these instances, New Yorkers were not necessarily the worst-behaved. Todd Martinez, who gathered helping data in both New York and Los Angeles, observed clear differences between the two cities: "I hated doing L.A. People there looked at me but just didn't seem to want to bother. For a few trials I was acting the hurt leg episode on a narrow sidewalk, with just enough space for a person to squeeze by. After I dropped my magazines, I remember one man who walked up very close to me, checked out the situation, then sidestepped me without a word. L.A. was the only city (of 12) that I worked where I found myself getting frustrated and angry when people didn't help. In New York, for some reason, I never took it personally. People looked like they were too busy to help. It was like they saw me but didn't really notice—not just me, but everything else around them."

To the stranger in need, of course, thoughts are often less critical than actions. Our data show that some people manage to find the time to help in fast places just as they do in slow ones. Rochester, the most helpful place of all, ranked a relatively fast ninth of thirty-six in our pace-of-life study. And the very slowest pace of life was found in Los Angeles—one of the least helpful cities (thirty-fourth). The bottom line is that your prospects are just as bleak in New York as they are in Los Angeles.

We are also finding the same on-again, off-again relationship between the pace of life and helping for cities on the international level. Over the past several years, my students and I have been testing variations of five of the helping measures used in our U.S. studies in a number of cities around the world.[19] Just as in our U.S. studies, some foreign cities support the hypothesis that a slow pace of life engenders social responsibility. We have found, for example, that people in Rio de Janeiro, who have a very slow pace of life, are extremely helpful to strangers on our measures. And people in Amsterdam, where there is a relatively fast pace of life, score low on helpfulness. But other places are clearly incompatible with the hypothesis. Bulgarians, for example, are very slow on our pace-of-life measures, but are no more helpful than fast-paced New Yorkers on the helping measures. And we have found that people in Copenhagen walk quickly and have a generally fast pace of life, but still receive high scores for their willingness to offer assistance to strangers.

These results make clear that even people with a fast pace of life are capable of finding time for others. And a slow pace of life is no guarantee that people will invest their saved time in practicing social ideals. In both fast and slow places, people either make the time to help or they don't.

EIGHT

JAPAN'S CONTRADICTION

The only way to go beyond work is through work. It is
not that work itself is valuable; we surmount work by
work. The real value of work lies in the strength of
self-denial.

KOBO ABE, *The Woman in the Dunes*

A lifetime of research has convinced me that, on the whole,
a rapid pace of life creates conditions conducive to both
higher life satisfaction and to a greater incidence of coronary heart disease. But in the quagmire of cross-cultural psychology, exceptions are sometimes more edifying than the rule. What
happens when a fast pace is embedded in a network of values that
buffer against stress? Japan offers a fascinating example.

WORKAHOLISM JAPANESE STYLE

The pace of life in Japan is one of the most demanding on earth.
As we have seen, not only do the Japanese work quickly, but they
work a lot. They avoid vacations and dread retirement. A supreme
reward for an outstanding employee, in fact, is exemption from
his company's mandatory retirement age.

Blue Mondays are no problem for Japanese workers. They are more likely to be afflicted with angst and psychosomatic symptoms from conditions with names like "Sunday Disease" (*Nichiyoy byou*) and "The Holiday Syndrome" (*Kyuujitu byou*). One doctor, for example, has described the case of a stricken accountant: "Every Friday, without fail, he feels a sharp pain spreading across the back of his neck. He spends the entire weekend in bed, too tired to move. But when Monday comes along, he is miraculously cured." The holiday syndrome, according to psychiatrist Toru Sekiya, "is a uniquely Japanese disease. These men can't stand not to be working."[1]

Perhaps most remarkably, the holiday syndrome is not necessarily seen as a pathology. Psychologist Isao Imai is a psychologist with Stress Management, a Japanese firm that consults with corporation managers about matters of employee stress. Imai says that his corporate clients encourage him to play down any "Type A" talk with the employees. "These managers tend to see workaholism as a goal rather than as a problem."[2]

The magnitude of Japanese dedication to work can be dazzling. One yardstick of this commitment may be seen in the genesis of government attempts to make people work less. In what must be an unparalleled Japanese project, government policymakers have for the past several years engaged in a formal campaign to get workers to slow down. I happened to be working at Sapporo Medical University in Japan during the summer of 1987 when the government first launched this "work less, play more" crusade. Over coffee one morning, I commented to a couple of my Japanese colleagues, both of whom were health psychologists, how encouraged they must feel that the government was finally acknowledging the psychological costs of overwork. They both did a doubletake at my *gaijin* naivete. The real motives, they explained, were economic. A little later one brought me an editorial from the *Asahi Evening News*, a Japanese English-language daily, which explained: "For the nation's domestic market to pick up, which the government says it must, a rise in consumer spending is a must. But for that, Japanese company workers will have to have more leisure time. Hence the Labor Ministry's work less, play more campaign."[3] So much for kicking back in Japan.

Whatever the government's motives, its campaign has been no match for the Japanese work drive. First, in its 1988 Economic Five-Year Plan, the government established the goal of reducing annual work time more than 20 percent, to 1,800 hours, by 1992. A new law was passed intended to gradually reduce the work week from 48 to 40 hours. Labor Ministry officials embarked on a 47-city lecture tour, delivering pep talks to workers on topics like "How to Work and Rest in a Relaxed Society" and "Our Company's Restful Week." All these efforts, however, ended in almost total failure. Follow-up surveys found that working hours were—and still remain—almost unchanged.[4]

Then the government decided to challenge the notorious Japanese aversion to taking vacations. As seen earlier, Japanese workers take advantage of only about half of the vacation days to which they are entitled. Ikuro Tagaki, a professor at Japan Women's University and an expert on the nation's overwork problem, comments that: "Taking holidays was nearly sinful when Japan was poor. Working hard was a universal virtue until very recently." Even the Japanese language reinforces the work ethic. The Japanese word for free time, *yoka*, literally translates as "time left." "Free time is just not regarded as equal in value [to working]," Tagaki says.[5]

The government, fully recognizing the steadfast attachment the Japanese have to their workplaces, chose to encourage vacations in a most Japanese way. Advertising campaigns blanketed the country with slogans such as: "To take a vacation is proof of your competence." One summer, posters went up across the nation showing an idyllic mountain scene with two relaxing Japanese in safari suits lying on the ground next to a reclining leopard. On each poster was the summer's slogan, "Hotto Week"—the Japanese word for relaxation and a pun on the English word "hot." Alongside that was a more direct message in Japanese from the Labor Ministry, which essentially translates as: "We order you to take one week of vacation." The *Asahi Shimbun*, a large Japanese newspaper, plastered its walls with a poster showing a ferocious-looking boss screaming into a telephone: "If you come to work, you're fired."[6] Once again, however, Japanese workaholism seems to be winning the battle. The rate of used vacation days has barely changed since the government's campaign began: 50 percent in 1986 versus 53 percent in 1992.

Recently, there has been a campaign in Japan encouraging work leave after the birth of a child. But the motives for these furloughs are very different from those behind the extended maternity leaves in countries like France and Sweden. In traditional Japanese fashion, the reasons have more to do with business and productivity than with concern about the mental health of women and their families. Mostly, they have been linked to the country's steadily declining fertility rates. In 1991, when rates hit a new low of 1.53, the government passed a law which offered both men and women a post-childbirth leave of absence from work. "Companies are suddenly worried they will not have enough workers," explained Sumiko Iwao, an expert on the role of women in Japan. Male lawmakers "almost felt as if the Japanese race was on the verge of extinction."[7]

Speed is a highly regarded virtue in traditional Japanese culture. It is said that the man who moves slowly is a fool. Wasting time, even for basic biological needs, is frowned upon. There is a saying (though not necessarily recited in the most refined circles): "*Hayameshi, hayaguso, geinouchi.*" Roughly translated it is: "To eat fast and defecate fast is an art."

A worker who moves too slowly, no matter whether the task actually requires speed, commits the ultimate sin in the Japanese workplace: not giving one's all. Garr Reynolds, an American who works for Sumitomo Electric in Japan, observes that the Japanese believe they "should appear to be busy at the office whether they are actually busy or not. One way to appear busy is to do things quickly; for example, jogging the 10 feet to the copy machine; pounding the keyboard of your PC as you compose a routine letter; bolting out of your chair ('Yes, sir!') every time your superior calls your name. To be and/or appear to be busy is a virtue in this society, and appearing to be doing things quickly, and with an element of panic, suggests to others that you are indeed busy and therefore a good employee."[8]

The Japanese can be so focused on the virtues of speed and hard work that serious cultural conflicts can occur. Some of the most tumultuous of these clashes have resulted from the remarks of Japanese leaders about the declining work ethic in the United States. Early in 1992, Yoshio Sakurauchi, the Speaker of the Lower House of Japan's Diet (Congress), commented publicly that "the root of America's [trade] problem lies in the inferior quality of American labor." Putting it even more squarely, Masao Kunihiro, an anthropologist who is also a member of the Diet's Upper House, stated around the same time: "Sadly, there's been an erosion of the Puritan work ethic in America, a country which taught us so much." These criticisms were met with an outburst of statistics-waving, Ben Franklinisms, and sometimes vicious Japan-bashing from government officials in Washington, mostly defending U.S. workers as underappreciated and misunderstood.[9] But to the Japanese, it seemed an obvious truth that if people aren't producing they must not be making the effort. The solution: just work faster and harder.

Japanese people are acutely aware of what hard work can achieve. In the quarter of a century following World War II, they watched their country's finances rise from the literal ashes of World War II to the current position of world prominence. They understand that there was little magic in their "economic miracle." It resulted from effort and personal sacrifice. The Japanese consider hard work and failure to be mutually exclusive.

BUFFERING AGAINST CORONARY HEART DISEASE

[The] surface cultural traits which appear so significant to us are like the carapace of the tortoise: they hide and protect the real Japan. Commodore Perry may have thought he "opened" Japan to the West; in fact, as with all cultures, what was revealed on the surface was little more than an illusion.

EDWARD HALL and MILDRED HALL,
Hidden Differences

Workaholism is the Japanese way of life. Should we conclude, then, that Japan is a nation on a collision course with coronary artery disease? Are the Japanese, as the American journalist Walter Winchell once said of himself, "go[ing] the pace that kills"?

Health statistics indicate a resounding "no." If anything, the opposite seems to be the case. Despite its apparently runaway pace of life, overall rates of death from coronary heart disease in Japan are remarkably low. They have the fifth lowest rate of coronary heart disease of the 26 countries in our study (CHD statistics were not available for five other countries in the study). In fact, Japan's coronary heart disease mortality rates were the very lowest of 27 industrialized countries compared in a report by the World Health Organization.[10]

How do the majority of workers in a fast-charging Type A population avoid coronary heart disease? Certainly, the low-cholesterol Japanese diet helps, but data indicate that it is not the whole answer. Cultural values appear to be an equally important buffer. The researchers Michael Marmot and Leonard Syme found that Japanese-American men who did not have a traditional Japanese cultural upbringing were 2 to 2.7 times more likely to experience coronary heart disease than those who had been raised in a more traditional household. This held true even when the traditional CHD risk factors—diet, smoking, cholesterol, blood pressure, triglycerides, obesity, glucose, and age—were taken into account.[11]

Japan's cultural values have at their core a profound and rigid focus on the welfare of the collective. Japanese-style workaholism is cast from a very different mold than its counterpart in the West. The traditional Japanese work ethic—"*senyu koraku* [struggle first, enjoy later]"—is framed within its collectivist value system. Hard work and productivity are not merely means toward providing for one's family, as is often the case in individualistic nations, but are one's civic obligation toward the betterment of the "tribe." In the United States, hard work and long hours have traditionally been adorned in terms of masculinity; they are the calling and responsibility of the family breadwinner. In Japan, these efforts are cast in dimensions of patriotism.

Japan's special blend of collectivism focuses on devotion to the group. For most workers, by far the most important of these groups is their company. Individualism is an alien notion to the Japanese worker, whose personal success is measured by the prosperity of the entire organization. Loyalty and devotion to the group is not an option in Japan, but a given. The novelist Yukio Mishima attempted to capture the intensity of this collectivist identification in his essay "Sun and Steel":

> The group was concerned with all those things that could never emerge from words—sweat, and tears, and cries of joy and pain. If one probed deeper still, it was concerned with the blood that words could never cause to flow ... only through the group, I realized—through sharing the suffering of the group—could the body reach that height of existence that the individual alone could never attain.[12]

Identification with one's company can be seen everywhere. Company employees begin each workday by standing to sing their company songs, reciting lines like "A bright heart overflowing with life linked together, Matsushita Electric." They often wear company colors to reflect their lifelong identification with their employers. The Japanese work for the success of their company, and then experience the company's successes as their own. "Your team can win even if you cannot" is a favorite slogan. As Suguru Sato, my former colleague at Sapporo Medical University, once explained to me, "I feel toward my department as you feel toward your family." Sato did not simply mean this metaphorically. Survey data indicate that 66 percent of Japanese workers—mostly males, in Japanese fashion—rate their companies as at least as important as their personal lives.[13]

In return for this devotion, workers have historically been assured the unconditional, permanent support and security of the group that they serve. Yoshiya Ariyoshi, formerly the chairman of Japan's biggest shipping line, once commented: "The very start of life is different for Americans and Japanese. In America, as I understand it, children are encouraged to assert their separate identity. Here in Japan, the first thing you learn is to harmonize with the group. In exchange for this conformity, people will be kind and considerate toward you. It's not necessary for you to ask for anything; your wishes will be granted without your asking. In childish terms, if you are good and don't make strident demands, people will spoil you."[14]

For most Japanese—especially men—the most important support group is their company. College graduates traditionally choose their employers with as much care as they do spouses. Like a marriage, the relationship with the company is intended to be for life. Much has been written about the lifelong financial and job security Japanese workers have traditionally enjoyed. But it is the emotional support that may matter most to them. When an employee gets sick, for example, not only does the company send over an in-house doctor, but it is common for the boss to pay a home visit to offer consolation. This spirit may have been carried to its extreme several years ago when officials of the Kyoto Ceramic Company purchased a mass tomb for its workers and their families. They wanted to be sure that their employees "won't feel lonely" after death.[15]

Especially for men, co-workers are not only the backdrop for the workday, but are nearly the totality of one's social world, making loneliness a rare problem for the Japanese employee. In fact, this emotional support may be another key to Japan's low rate of coronary heart disease. Health psychologists offer considerable evidence that a sense of social support is a powerful buffer against stress and illness. Studies conducted in a number of countries have found that a strong support system lowers the likelihood of many illnesses, decreases the length of recoveries and reduces the probability of mortality from serious diseases. It has been demonstrated, for example, that people with higher levels of support recover faster from kidney disease, childhood leukemia, and strokes, have better diabetes control, experience less pain from arthritis, and live longer.[16] Most significant for our research, studies show that they also have a lower probability of having a heart attack and are less likely to die from one. Redford Williams of Duke University, in a major study of patients with advanced coronary heart disease, found that 82 percent of those with extensive support networks survived at least five years, as opposed to only 50 percent of those who were most socially isolated.[17]

Besides the emotional support the company provides, the Japanese worker benefits from knowing that colleagues are all carrying their share of the load. By working for and with the group, the pressure is divided among one's co-workers. As a result, Japanese workers appear to avoid much of the stress that typically accompanies hard work in the West.

This theory is supported by cross-cultural studies measuring Type A behavior. The Jenkins Activity Survey, the most widely used measure of Type A behavior, includes a series of questions about being hard-driving—competitive, short-tempered, impatient—as well as a series about being hard-working. For people in the United States (where the scale was developed), high scores on the first set of questions usually go hand-in-hand with high scores on the second: both are part of the same behavior pattern. But in Japan, where social harmony is the most highly respected social value, competition and aggression have little place. Studies show that when Japanese take the Type A personality scale their answers on questions reflecting hard work bear little relationship to their answers on items measuring hard-driving behavior. They tend to score high on hard work but low on hard-driving behavior.[18]

Even translating the Type A questions about being hard-driving is a problem. One frustrated researcher reported that the best Japanese translation he could find for the question "Do you like competition on your job?" was "Do you like impoliteness on your job?"[19] Sato comments on this problem of translating words like "competition," "assertiveness," and "aggression" into Japanese: "In Japanese, the word 'aggression' translates as '*kougeki sei*,' which literally means 'attack.' When a person is described as *kougeki sei*, it means that he is hostile, or evil-tempered. It has a very negative connotation." In the West, being aggressive can also take on a positive connotation, as in "assertiveness." But assertiveness is a difficult word to translate into Japanese. Whereas Western children are taught, "Ask and you shall receive," the properly socialized Japanese believes that it is when you are silent that things will come to you.

Competitive hostility and anger, it appears, play little part in the hardworking, fast pace of the Japanese. But in the United States and other Western cultures, where our studies observed a strong relationship between a rapid pace of life and coronary heart disease, there is often a fine line between speed and time urgency on the one hand and competition and hostility on the other.

The case of Japan suggests that time urgency is not a direct cause of coronary heart disease. It is only when speed and temporal pressures co-exist with the toxic elements of hostility and anger that they show a substantial relationship to coronary heart disease.[20] But Japan's low rates of heart disease demonstrate that speed and time urgency need not be lethal in themselves. This is good news for workaholic Westerners who resist their cardiologists' warnings to slow down. As long as work is approached with the right attitude—without hostility or competitiveness—there appears to be little or no increased risk for coronary heart disease.

KNOWING YOUR *GIRI*

The balance and temporal flexibility that characterize Japanese workaholism have roots in the principle of *giri*, or obligation to others. The rules guiding appropriate social behavior are tightly choreographed in Japan. Virtually every social relationship is structured around clearly delineated duties: one's *giri*. Sometimes the expectations can pertain to the trivial; for example, people talk of a *"girichoco"* at Valentine's Day—their "chocolate obligation." In fact, there is so much *giri* in Japan that there are second-hand gift stores, dedicated to recycling all the presents that are exchanged.

But the concept of *giri* goes much deeper than present-giving. At the heart of one's duties are carefully prescribed obligations to one's family, company, and country. It is this level of *giri* that powers the group focus defining Japanese workaholism.

Allen Miller, an Australian who has taught English in Nagoya, speaks about the pride that the Japanese feel when successfully fulfilling their *giri*. He points out how the Japanese like to have everything well planned out. "The Japanese I've met often become almost elated when they understand exactly what their obligation is to me," says Miller. "The important issue is not how much of an obligation they have, but to understand exactly what is expected of them. They're then happy to go about fulfilling that obligation."[21]

The *giri* behind their work ethic has allowed the Japanese to become masters at switching from one temporal mode to another as the situation requires. They are second to none when the situation calls for an up-tempo. But their *giri* in the workplace sometimes calls for slowing down the tempo, and the Japanese workaholic can often perform at this cadence with equal skill.

The Japanese mixture of temporal modes reflects their fundamental attitudes toward the nature of work and nonwork time. In the United States, there is a sharp distinction between the time for work and the time for play. One's commitment to the company begins and ends at specific times, often calibrated to the minute. The employee is expected, within clearly defined limits, to offer single-minded focus on the task of the day. The employer, on the other hand, understands that the worker's after-hours time is, unless special arrangements are made, off-limits to the company.

For Japanese workers—again, male workers in particular—there is a fuzzy line between the times of work and social life. To most Japanese men, their co-workers and friends are one and the same. Japanese workers' hours are long, but their production is not as great as Westerners might expect. They spend a lot of time talking with colleagues, going to meetings, idly chatting. Their workday, in other words, is not just production-oriented. Suguru Sato, the junior member of his faculty, describes how one of the *giris* of his job, in fact, is to be present while his senior colleagues are having their morning coffee, their lunch, or even playing board games with one another—even though he may have other work begging for his attention, which he will need to stay overtime to finish. Social down time, he explained, is necessary for the *wa* (harmony) that he and his colleagues, and Japanese society in general, all value so highly. As such, he perceives those things that a U.S. professor might consider wasting time to be a very important part of his job. "When you're hired by a company for life," he observes, "harmony becomes critical."[22]

The Japanese, being so group oriented, feel much less need for private time than do people in the United States. Even after a long day on the job, workers at Sapporo Medical University will often hang around for another couple of hours, perhaps drinking beer and watching a ball game with their co-workers. These habits help create the desired harmony in social relations. And the resulting *wa* feeds the feeling of personal responsibility and intrinsic motivation that drives the productivity of the Japanese workforce.

This acceptance of one's *giri*, and the willingness with which people then take on whatever is expected, no matter how much work or time it may require, underlies the Japanese dedication to the company. Everyone understands that their *giri* is important—that the happiness of their family, the success of their company, and the future of Japan is determined by how each individual does their job. In part, this results from Japan's seniority system, whereby nearly every eventual leader begins with the most menial job in the company. The president of the bus company began by driving a bus; the manager of the restaurant started out as a waiter. Every worker is part of the same operation, and each understands that the lowest blue-collar job is essential to the success of the operation.

Allen Miller talks about how, in his classes, "I couldn't get people to understand what it meant when, after a worker phoned in sick in the United States, a supportive boss would answer 'Don't worry. You just get your rest. We'll be fine without you.' The students simply couldn't understand. In Japan, the answer is something like 'Oh, we'll do our best to carry on without you, but it will be very difficult.'"[23]

The *giri* principle helps explain the Japanese ability to work hard without killing themselves. Japanese workers who put in long hours understand that this is their *giri.* All of their obligations are well planned out, and so long as these obligations are met, seniority will bring just rewards. It is this belief—that what they are doing is important, that it is appreciated by those above them, and that it is part of a group effort—that many Japanese believe to be their principal buffer against stress.

KAROSHI, OR DEATH BY OVERWORK

This is not to say that Japanese workaholism is without costs. In fact, more and more Japanese have begun to fear that compulsive duty to the workplace has overshot the mark—that it is no longer in the best interests of either individuals or the tribe. Hard work, these critics argue, has evolved into an addiction, and it is exacting a steep price from the well-being of many people and their families.

Few individuals are more in touch with the price of Japanese-style workaholism than Tokyo lawyer Hiroshi Kawahito, who heads the Karoshi Hotline. The word "*karoshi*" refers to death by overwork, usually from coronary heart disease. Not only does the hotline take calls from employees suffering work-related illnesses, but it counsels families who are suffering from the loss of their husbands and fathers to company overtime. The first Karoshi Hotline, which opened in Osaka in 1988, received 309 calls in its initial day of operation. A little more than one year later, there were Karoshi Hotlines in 28 prefectures. The number of karoshi cases, Kawahito says, continues to increase. "At first there were karoshi hotlines in seven locations but we sometimes got calls on the Tokyo hotline from people as far away as Okinawa, so we thought we should expand the number of hotline offices," he says. "I'm now working to get the number of karoshi hotline centers up to 47—one for each prefecture." It might be noted that Kawahito himself works ten hours per day in his efforts to keep people from working themselves to death.[24]

Clearly, many Japanese are paying a price for their workaholism. The Karoshi Hotline may be a signal that the psychological buffer of the group is wearing thin for many Japanese workers. And there is reason to fear that stress-related problems will become increasingly common in Japan as economic pressures begin to threaten the previously untouchable elixirs of the seniority system and lifelong job security.

Taken as a whole, however, the evidence continues to be strong for the psychological resiliency of the Japanese worker. Most Japanese appear to be relatively comfortable with their regime of hard work—compared, at least, to their colleagues in the West. On the surface, the Japanese norm fits very neatly under the Western label of workaholism. If this is workaholism, however, it is a very different breed of it.

LEARNING FROM JAPAN

It has been said that Japan is the ideal place from which to observe the rest of the world. One can see why, for perched out there on the extreme edge of Asia it often seems as if one is looking at the world from the outside.

IAN BURUMA, *Behind the Mask*

In Japan, it is said, there is always the contradiction. The Japanese apologize to no country when it comes to speed, but they are not necessarily tyrannized by the clock. Writer Pico Iyer, who spent a year in Kyoto, observed that the Japanese are "connoisseurs" of time. They "package Time and turn the bumpy chaos of successive moments into an elegy as beautiful as art." They are able to excel as much at slowness as at speed. "So much of Japan," Iyer felt, "[is] set up as a retreat from Time, a way to stay Time, or step out of it."[25]

Perhaps the most important message from Japan is what it teaches about ourselves. People in the West have often come to view the choice between rushing and leisurely activity as a tradeoff between accomplishment on the one hand and stress on the other. There is no question that hard work may, in fact, all too often exact a steep price on contemporary workers, but as the Japanese experience demonstrates, this relationship is not universal.

Recent Type A researchers have argued that behaviors such as moving fast, talking fast and being involved in one's job do not necessarily cause coronary problems as long as they are pursued without hard-driving competitiveness and hostility.[26] Given that these conclusions are primarily based on data from the United States, which is where most Type A research has been conducted, it is paradoxical that some of the strongest support for this case comes from the edge of Asia. Even though results from our pace-of-life studies indicate that in many countries, the United States and much of Western Europe in particular, time urgency may be more often associated with coronary heart disease than many current Type A researchers believe, the Japanese example suggests that this relationship is not inevitable: the need to make every second count, to work hard and move at a rapid pace need not necessarily be destructive to one's health.

The use of time is yet another instance where the Japanese have borrowed some of the more attractive features of Western culture but, at a deeper level, held fast to their own traditional values. In recent years, Western businesspeople have begun humbly admitting that it is now their turn to learn production techniques from the Japanese. So, too, they might take a lesson from the Japanese about controlling time.

PART III

CHANGING PACE

NINE

TIME LITERACY

Learning the Silent Language

Having accepted an invitation, it is wise to observe
the rules. Punctuality is a must, as there is not often
any margin for cocktails. If the dinner invitation is
for seven o'clock, you press the button in the host's
apartment at precisely seven and in a few minutes
you will be seated at the table. If a taxi brought you to
the door ahead of time, you wait downstairs. Perhaps
other guests have gathered there before you, and not
until the cathedral clock in the neighborhood strikes
seven do you proceed.

LILLY LORENZEN, *Of Swedish Ways*

The historian Lewis Mumford once observed how "each culture believes that every other space and time is an approxi-
mation to or perversion of the real space and time in which
it lives."[1] But the truth of the matter is that there are no overriding
rights and wrongs to a particular pace of life. There are simply *different* ways of life, each with their pluses and minuses. All cultures,
then, have something to learn from others' conceptions of time.

But accessing the temporal codes of other cultures requires effort. Temporal patterns are at the crossroads of a vast web of cultural characteristics; they permeate the personality of a place. As the case of Japan makes clear, their psychological meaning to insiders cannot be properly understood in isolation from this broader context. One needs to understand the fundamental values of a culture before coming to terms with its time sense. No wonder outsiders become befuddled when trying to comprehend this silent language.

In many instances, temporal illiteracy leads to situations that are simply awkward and embarrassing; in other cases, however, the lack of knowledge can be socially disabling. The latter is often the result when non–clock-time people must achieve by the standards of fast-paced cultures. There are entire subpopulations within otherwise economically vital communities who are marginalized by their inability to master the clock-governed pace of the mainstream culture. These temporally disabled subgroups are particularly common in societies with large multiethnic, multicultural populations, especially those that are undergoing rapid social change. Their temporal perspective is often limited to the present moment. The social critic Jeremy Rifkin argues, in fact, that temporal deprivation is a built-in feature of all advanced societies. "In industrial cultures, the poor are temporally poor as well as materially poor," says Rifkin. "Indeed, time deprivation and material deprivation condition each other . . . those who are most present oriented are swept into the future that others have laid out for them."[2] Edward Banfield, in his book on urban poverty, *The Unheavenly City*, goes even further: "Extreme present-orientedness, not lack of income or wealth, is the principal cause of poverty in the sense of 'the culture of poverty.'"[3]

There are areas in the United States that offer vivid evidence for these assertions. Dolores Norton, a professor in the School of Social Service Administration at the University of Chicago, has been studying the intellectual development of children in poor U.S. families for more than a decade. Her research has focused on the experiences of an extremely high-risk sample—the children of low-income black adolescent mothers living in the most blighted, impoverished pockets of Chicago. Norton videotapes the social interactions of these children at home and then uses this information to pinpoint the most difficult barriers they face in school. She has found ample evidence for Rifkin's and Banfield's temporal deprivation theories, seeing over and over again how many of the children's most formidable problems relate to an unpreparedness for the mainstream culture's use of time.

Temporal conflicts are almost unavoidable for these children, she finds, because their home lives are almost devoid of references to time. Daily routines, like parents leaving for work or having meals at regular hours, are rarely experienced by children whose parents are more concerned with problems like avoiding drugs and gang violence and trying to find food to put on the table. Most of these children's parents rarely provide instructions like, "Clean up your room before you watch your favorite TV show at eleven," or even simple sequential directions such as "First put on your socks and then your shoes."[4]

When the children enter school, they are often mystified by its temporal expectations. Norton describes this scenario:

Imagine yourself as a child in a classroom with adults who speak your language but whose directions you are unable to interpret, even though you may wish to please them. When you get up to see the gerbils, you are told to sit down, finish your coloring and wait to see the gerbils during free time. When you sit down to color, your paper is taken away before you finish, because it's ten o'clock and time for juice. Before you finish juice, it's "potty time."[5]

Norton has discovered that these types of situations are the norm for the inner-city children she has studied. The children watch their recess game halted in midstream because it is now snack time and then find their unfinished snack taken away because a new lesson must begin. The less congruent their concept of time is with that in the classroom, the poorer their achievement. The children's confusion and frustration often lead them to rebel or withdraw. They may become labeled as troublemakers or as slow learners. And so the downward spiral proceeds.

Mastering another culture's silent language, as Norton's discouraging account makes clear, is a formidable and treacherous pursuit. These difficulties in adjusting to another time sense are, of course, not limited to economically deprived subcultures; in earlier chapters we have seen how precipitously and profoundly some of the most international of people, from politicians and kings to cross-cultural psychologists, have stumbled badly over the temporal rules of other groups. But there are also multitemporal success stories.

One group that has demonstrated a proficiency for temporal flexibility is the thousands of Mexicans who live in Tijuana but commute daily to jobs on the California side of the border. Psychologist Vicente Lopez, currently director of the library and an instructor in the Communications Department at the University of Mayab in Merida, Mexico, considers himself typical (temporally, at least) of this group. Lopez spent five years making the Tijuana-to-San Diego commute. He talks about how each time he crossed the border it felt like a button was pushed inside him. When entering the United States, he would feel his whole being switch to rapid clock-time mode: he would walk faster, drive faster, talk faster, meet deadlines. When returning home, his body would relax and slow into event time the moment he saw the Mexican customs agent. "There is a large group of people like me who move back and forth between the times," Lopez observes. Many, he believes, insist on keeping their homes on the Mexican side precisely because of its slower pace of life. "In Mexico, we are inside the time. We don't control time. We live *with* the time. You have to tell the Americans, 'Please understand how to act on Mexican time.' Then you have to tell Mexicans, 'Please understand that Americans are like this.' And then you can switch back and forth between the two different times."[6] Lopez says that this switch has become second nature to many of the Tijuana-San Diego commuters, himself included.

Vicente Lopez proves that unfamiliar patterns of time can be mastered. Of course, most intercultural travelers would prefer to avoid the five years of on-site mistakes that Lopez had to endure before achieving multitemporal proficiency. To simplify the process, might it be possible to formally teach the fundamentals of another culture's time sense, in the same way people are taught the spoken language? This is, in fact, exactly what Dolores Norton has begun to accomplish with her Chicago group.

Her temporal training program is not unique. In Israel—perhaps the one country today that can match the demographic diversity of the United States—psychologists Ephraim Ben-Baruch, Zipora Melitz and their colleagues at the University of Ben-Gurion in the Negev have reported success with an elaborate set of time-teaching exercises they have designed to train children from third-world cultures to adapt to Israel's mainstream pace of life. Their program consists of 29 wide-ranging activities that teach 8 basic concepts of time. Many of the topics read like a remedial course in Piagetian temporal development. For example, one group of activities teaches the concept of "before and after." In this module children learn concepts ranging from how to prepare an attendance card to understanding how the mainstream culture distinguishes between concepts of past, present, and future. Another module, "day and night," teaches the mainstream culture's accepted standards of the parts of the day and night (morning, noon, evening, night) and the sequence of the sorts of activities that occur during these time frames. A third module teaches the days of the week and the different expectations for each day.

Other modules focus on less tangible temporal concepts. During one series of activities, children are presented with the concept of duration. For example, children perform everyday activities like buttoning their shirts and lacing their shoes and are then asked to estimate how long each took and which was longer. They are later presented with longer temporal durations: How much time does it take seeds to sprout? To go up ladders and down slides? How long to grow up? Later exercises teach the concept of simultaneity—that some events occur one after the other, but that other things can happen at the same time. The last three modules focus on the different patterns of time flow: cyclic time, linear time, and understanding the notion of time as a limited entity. To explain the concept of cyclic time, children are taught, for example, the cycles of holidays and of nature and how these events recur in the same sequence over and over again. By grasping time as cyclical, they also come to understand it as reversible: spring will return, there will be other birthday parties, tomorrow is a new day. Children are next asked to understand the less "natural" concept of linear time. This is usually a more difficult concept for them, many having come from traditional Bedouin upbringings or from some other desert or village background. To understand linear time, they study the types of events that tend to be unrepeated, that "flow in one direction." They are taught how the linear concept of time is related to an understanding of beginnings, durations, and endings. In the last module of exercises children learn the concept of time as a limited entity.[7]

Norton's and Ben-Baruch's syllabi both teach children the sometimes ugly reality that many tasks in the dominant culture have an allotted and limited time in which they may be completed. By preparing children to deal with ideas like time limitations, the value of time, and the desirability of efficiency, they are helping them to understand that in their new culture anyone who fails to master the clock may be labeled a failure.

EIGHT LESSONS

The programs devised by Norton and Ben-Baruch target event-time people who are preparing for encounters with faster cultures. But if you are moving in the opposite temporal direction, from fast to slow, there is as much or more to learn from temporal training. What lessons can we offer a sojourner from a "time is money" culture, like the United States, to help them adapt to the time sense of Vicente Lopez's Mexico? And what if you have no intention of physically leaving your own culture, but would simply like to enhance your temporal repertoire, to learn alternative approaches to time that may lead to a more satisfying life? Can we assign words to the silent language? Here are a few lessons that can be learned by clock timers who wish to understand the temporal logic of slower cultures.[8]

Lesson One

Punctuality: Learn to translate appointment times. What is the appropriate time to arrive for an appointment with a professor? With a government official? For a party? When should you expect others to show up, if at all? When a class is scheduled to start at 10:00, at what time should the instructor begin marking students late? How much importance should be assigned to arriving late? What sort of apologies or excuses are expected and acceptable from late arrivers? How promptly should the class end? Is there a social message in arriving late (a real big shot?) or promptly (a nobody? "Did you arrive with the clean-up crew?" as the taunt goes in Mexico)? Should we expect our hosts to be upset if we arrive late, or promptly? Are people expected to assume responsibility for their lateness?

Many of these cultural rules can be taught. Sojourners should seek guidance about the expected ranges of promptness, if any, for the sorts of situations they are likely to encounter. You can learn to translate *hora ingles* into time frames like *hora mexicano*, *hora brasileiro*, Indian Time, CPT, and rubber time. You can be prepared beforehand for the sort of critical situations that are likely to occur when your conceptions of punctuality are at odds with those of your hosts. Say, for example, that you are assigned to work in Saudi Arabia. The first day of the job you arrive full of anticipation. Gloom quickly sets in, however, when your first appointments fail to show up after half an hour. Have you been rejected? Before packing your bags for home, it would save you a lot of trouble to understand that thirty minutes has a very different meaning in Saudi Arabia than it does back home. In the United States, the major unit of time for assessing punctuality for appointments is usually on the order of five minutes. When visiting traditional Arab cultures, however, you should know that the corresponding unit of time is fifteen minutes. If an Arab is thirty minutes late by your clock, he is only ten minutes late by his own standards. You should be willing to wait a half-hour or more for your host, or *he* will be insulted.[9]

You can also be taught a culture's customs for making and keeping appointments. The fundamental cultural clash here often comes down to what is more important: accurate information and facts or people's feelings? I, for one, have not fared particularly well with this lesson, and have paid for it. Before accepting my position in Brazil, for example, I asked my prospective employer if she could also find work for a friend who wanted to come along. "*Não tem problema*," my employer answered—a response that was to become for me a familiar but meaningless refrain. After our arrival, we inquired again about my friend's job. My *chefe* made a phone call, scribbled out a name and address, and told us to be outside her building the next day at 9 A.M., *en punto* (sharp), at which time her chauffeur would take us to the interview. Pretty impressive, we thought. Except that the driver never showed up. The same scene was replayed the next day. We were furious with my *chefe*, even more so when she failed to apologize. Deciding to take matters into our own hands, we began asking other seemingly well-connected people if they could find work for my friend. In every case, they quickly and enthusiastically assured us that they knew just the person to talk to. *Não tem problema!* Without fail, they told us that their chauffeur would take us there. And not one of those drivers ever showed up. After about five of these misfires, I altered my tactic. The next time I received a "yes" to my request, with the usual offer of a driver, I explained with that charming U.S. forthrightness about how it was O.K. if they couldn't help, but it would really be so much nicer to get an up-front "no" than to have to wait all the next morning for a no-show driver. My colleague looked uncomfortable and insulted. "But Dr. Levine," he responded, "I'm telling you that I know just the person to help your friend." He then belted out one of the most voracious *Não tem problema's* I had yet heard. And then the usual: no driver, no job, no apology.

When we described our frustration to a Brazilian friend, he explained *my* error, one I was to repeat many times: how a "yes" often meant "no" and that it was more important to Brazilians to appear helpful and polite than to stand by their time commitments. My friend then scolded me for having backed my colleagues into a corner by making a request they could neither deny nor deliver on. Rejecting my request would have been both rude and an acknowledgment of their inefficacy. Missing an appointment is simply a severe case of lateness, a well-accepted Brazilian behavior. And in Brazil, people's feelings are more important than accurate information.

Richard Brislin tells of a similar experience with a Japanese friend: "Oftentimes a 'Yes' is just a way of saying, 'You've really got a good idea there.' One of my colleagues is a Japanese national . . . I [might say to her]: 'Work meeting, noon, Friday at the Maple Garden Restaurant. Okay?' She says, 'Maple Garden, what a great choice for a restaurant.' Now is that a 'Yes' or a 'No'? It's a 'Yes' from my perspective. But from her perspective, she's just telling me what great taste I have in restaurant choices. She hasn't committed to showing up. Oftentimes 'yes' meaning 'no' and 'no' meaning 'yes' is to keep a positive current going through the relationship."[10]

Misunderstanding each other's silent messages often leads to culture clashes where well-intentioned behaviors are attributed to shortcomings of character. Western businessmen, for example, have too often concluded that their Japanese counterparts can't be trusted; that they are two-faced, dishonest, and untrustworthy people who will make promises at the meeting table but not come through later. The Japanese, on the other hand, tend to interpret the accusation that they have lied as proof of Westerners' social insensitivity: the Westerner is either too lazy or incapable of reading the meaning of yes, no, and silences within the social context.

Many of these problems can be avoided. People who are knowledgeable about the target culture can lay out the norms of punctuality and appointment making for as many situations as possible, and also try to explain the logic behind these rules through the eyes of an insider.

Lesson Two

Understand the line between work time and social time. What is the relationship between work time and down time? There are easy answers to some questions: How many hours are there in the work day? The work week? Is it five days on followed by two days of rest? Or six to one, or four-and-a-half to two-and-a-half, or what? How many days are set aside for vacation, and how are they spaced?

Other questions are more difficult to get a handle on. For example, how much of the work day is spent on-task and how much time is spent socializing, chatting, and being pleasant? For Americans in a big city, the typical ratio is in the neighborhood of about 80:20—about 80 percent of work time is spent on-task and about 20 percent is used for fraternizing, chitchatting, and the like. But many countries deviate sharply from this formula. In countries like India and Nepal, for example, be prepared for a balance closer to 50:50.[11]

When you are in Japan, the distinction between work and social time can often be meaningless. The workday there has a large social element and social time is very much a part of work. The crucial goal that overrides both of these types of time is the *wa* of the work group. As a result, dedicated Japanese workers understand that having tea with their peers in the middle of a busy day, or staying overtime to down a few beers and watch a ball game, is an essential and productive part of their jobs. Private time, which many Americans believe is part of their Bill of Rights, is not terribly important to the Japanese. Want a good laugh? Try explaining to a Japanese company man about the contemporary American demand for guaranteed leave for "personal days."

The outsider not acquainted with a culture's norms concerning the balance between work and social time, or who is not prepared to go along with them, will quickly find himself or herself isolated.

Lesson Three

Study the rules of the waiting game. When you arrive in a foreign culture, be sure to inquire about the specifics of their version of the waiting game. Are their rules based on the principle that time is money? Who is expected to wait for whom, under what circumstances, and for how long? Are some players exempt from waiting? When and where is the Siddhartha move a viable alternative? What is the protocol for waiting in line? Is it an orderly procedure or, as in India, do people just nudge their way through the crowd, pushing the people ahead of them, until they somehow make their way to the front? Is there a procedure for buying oneself a place in front, or off the line completely? What social message is being sent when the accepted rules are broken? Either you learn these rules or you are condemned to plow, like a foreign water buffalo, through your hosts' temporal landscape.

Lesson Four

Learn to reinterpret "doing nothing." How do your hosts treat pauses, silences, or doing nothing at all? Is appearing chronically busy a quality to be admired or to be pitied? Is doing nothing wasted time? Is constant activity seen as an even bigger waste of time? Is there even a word or concept for wasted time? Of nothing happening? Doing nothing? What must it be like to live in a country like Brunei, where people begin their day by asking "What isn't going to happen today?"

You may have the opportunity to discover how curiously relaxing it can be to sit together in silence, free from plans, simply waiting for what happens next; and to eventually gain the reassurance that something always does. From the Japanese, you can learn how the spaces between events are as significant as the events themselves. People in the West are mostly attuned to the arrangement of objects; in Japan, it is the arrangement of spaces—the *ma*, or intervals—that are the focus. *Ma* teaches that the stops along the way are as meaningful as the eventual destination. The traditional

Japanese garden, for example, is designed with stepping stones that require the observer to stop and look down and then up again; as a result, every step offers a different perspective.[12] As the Westerner Artur Schnabel observed about his own art, "The pauses between the notes—ah, that is where the art resides."[13]

Mustn't all significant action be preceded by an incubation period? If you are going to China, you will find that the waiting period is not just a delay to be endured in order to reach the right moment. It is respected as the very creator of that moment.

Lesson Five

Ask about accepted sequences. Each culture sets rules about the sequence of events. Is it work before play, or vice versa? Do people take all of their sleep at night, or is there a siesta in the midafternoon? Is one expected to have coffee or tea and socialize before getting down to serious business, and if so, for how long? There are also customs about sequences over the long run: for example, how long is the socially accepted period of childhood, if it exists at all, and when is it time to assume the responsibilities of adulthood?

Misunderstandings about accepted sequences can land the visitor in serious trouble. One particularly vulnerable and volatile area, for example, is love and intimacy. Even in one's own culture, it is difficult to move smoothly through the intimacy cycle; when crossing cultural boundaries, it becomes a minefield. When is it time to move from one romantic stage to another? At what point does a couple progress from dating to an exclusive relationship? Often cues do not translate across borders. In the United States, for example, many young women believe there needs to be a physical event before a relationship is exclusive. The event doesn't have to be sexual intercourse, but there needs to be at least some touching, holding hands, certainly long passionate kisses. But in many other cultures—Japan, Israel, and Taiwan, for example—a physical event is not necessary. Consequently, a young American woman who takes up a nonphysical relationship with a man in one of these countries will be prone to assume it is just a nice platonic or professional affair. To her surprise, however, the woman hears that the whole town is gossiping that the man and she are an item. In her thinking, it cannot be time because there hasn't been a physical event. But to someone from another culture, a physical event isn't necessary for interpreting that the relationship has moved on.[14]

Should you fall into a long-term intimate relationship, be aware that the expected sequences may be troublesome. In the United States, the vast majority of people believe that romantic love— "true" love, of course—is an essential prerequisite to marriage. But this is hardly the case worldwide. In a recent survey, my colleagues and I asked respondents in eleven countries: "If a man [woman] had all the other qualities you desired, would you marry this person if you were not in love with him [her]?" In the United States, approximately 80 percent of both men and women answered flat-out "no." The percentage of people who ruled out marriage without love was considerably lower, however, in other countries: only 24 percent in India, 34 percent in Thailand, and 39 percent in Pakistan, for example. Looking at the eleven countries as a whole, we found that people from individualistic (versus collectivistic) cultures were much more likely to believe that romantic love should precede marriage. This was also true for people from economically well-off countries. It may be no coincidence that these two characteristics—individualism and economic vitality—were also related to a faster pace of life in our 31-country study.[15]

Most cultures believe in some type of romantic love. In the majority of cases, however, it is assumed that love is something that develops *after* the commitment of marriage, rather than the other way around. After all, how can you really love someone before learning what they are like to live with? And isn't it a bit ridiculous to make a lifelong arrangement based on an emotional reaction? The assumption that love comes after marriage is particularly common in cultures where marriage decisions are traditionally based on economic arrangements between families: the vast majority of cultures in the world. In a study of the marriage patterns of 850 distinct societies, anthropologists Erika Bourguignon and Lenora Greenbaum[16] found that some form of bride-price or dowry or other type of financial exchange exists in more than 70 percent of these groups. (In many countries, these arrangements have simply become more expensive. In Libya, for example, fathers used to demand a fee of about $3,500 in cash, with a camel, sheep, and a few gold coins thrown in, for marrying off their daughters. With Libya's oil boom, however, it is now not unusual for the groom's family to plunk down gifts of $35,000 or more, raising fears in some quarters that many Libyan women are being priced out of the marriage market.)

When is it time for marriage? To fall in love? It depends on where you are. One thing for sure is that the intelligent visitor had best learn some answers before getting in too deeply.

Before leaving the "When is it time for . . . ?" lesson, one other nasty cultural misunderstanding should be mentioned: the time it takes to move from out-group to in-group status. How long should you expect to be an outsider? You may find yourself being treated pleasantly enough, but you may still be frustrated by your hosts' unwillingness to reach out more closely. What is most important in this lesson is to recognize that cultures vary in their modal time for in-group acceptance. In parts of the United States that are used to heavy migration the waiting period will be considerably smaller than it is in closely knit cultures like Japan, where many foreigners perceive that the outsider's status is unalterably permanent. (The Japanese word for foreigner, "*gaijin*," is literally translated as "outsider." Even from a legal standpoint, it has been nearly impossible for any immigrants—other than perhaps a few famous sumo wrestlers—to ever receive Japanese citizenship.) Be prepared for what time frame to expect.

Lesson Six

Are people on clock time or event time? This may be the most slippery lesson of all. For the first five lessons, there are aspects of a culture's rules that can be translated relatively concretely: the accepted range of punctuality for a particular situation; the percentage of the work day that is spent socializing; who is expected to wait for whom; the length of time a silence must be endured before a "yes" means "no"; and even many of the cues that signify to the outsider when it is time for something to happen. But a move from clock time to event time will require a complete shift of consciousness. It entails the suspension of industrialized society's temporal golden rule: Time is Money. For most of us who have been socialized under this formula, the shift requires a considerable leap.

Nonetheless, outsiders can learn some of the behaviors that might be expected in event time cultures. Richard Brislin, for example, describes a common scenario faced by visiting professors: "Imagine you have an appointment at 11:30 with a very solid student who always turns their assignments in on time and is really working along their degree program. They're going to get out on time. There's another student, though, who's been slow to come up with a thesis topic and is getting an awful lot of Bs and Cs even in the graduate program . . . At 11:25 this student comes and says, 'Professor, I finally have a thought, I finally have a potential thesis topic.' Who has more claims on our time, the student who had the 11:30 appointment or this student who shows up at 11:25? Who has claims on our time?"

In clock time cultures like the United States the priority obviously goes to the student with the appointment. But if you find yourself in an event time culture, be prepared for the drop-in to expect the temporal right of way. "I'm very happy to work in cultures with event time," Brislin says. "I just hope someone tells me which one (event or clock time) we're on. That's all I ask them to do."[17]

The same sort of lessons apply to your expectations when moving from monochronic cultures, where one activity is scheduled at a time, to polychronic cultures, where people prefer to switch back and forth from one activity to another. In a polychronic culture, don't be insulted when your hosts become distracted from their business with you. It's simply cultural expectations talking. Nothing personal. To polychronic insiders, in fact, it would be rude *not* to shift their attention when the unexpected drops in. Moreover, you, too, will be expected to exercise polychronic flexibility, or else expect to be condemned as a social boor, a poor team player, and an inefficient worker.

Lesson Seven

Practice. An intellectual understanding of temporal norms does not in itself insure a successful transition. You can memorize other people's rules about time all you want but may still be totally dysfunctional when confronted with the real thing. As they say in the city, "He can talk the talk, but can he walk the walk?" The well-prepared visitor should seek out homework assignments that utilize on-site practice. Innovative teachers have been known to devise rather elaborate practice conditions. Anthropologist Greg Trifonovich of the East-West Center, for example, used to prepare Peace Corps volunteers and teachers for conditions in rural Pacific societies by creating a simulated village. One of the behaviors that Trifonovich taught was how to live without clock time. He showed students, for example, how to tell time by observing the sun and the tides.[18] Whatever your technique, realize that mastering the language of time will require rehearsal, and mistakes.

But be assured that it is well worth the effort. Cross-cultural training produces a wide range of positive skills. Research has shown, for example, that people who are well prepared for transcultural encounters have better working relationships with people from mixed cultural backgrounds; are better at setting and working toward realistic goals in other cultures; are better at understanding and solving the problems they may confront; and are more successful at their jobs in other cultures. They also report more pleasurable relationships with their hosts, both during work and free time; are more at ease in intercultural settings; and are more likely to enjoy their overseas assignments. The most astute of cross-cultural students also seem to develop a more general interest and concern about life and events in different countries—what has been called a general "world-mindedness."[19]

Lesson Eight

Don't criticize what you don't understand. Last, a guideline about observing cultures in general: the trap most difficult for the student of culture to avoid or escape concerns the inference of meaning. Almost by definition, cultural behaviors signify something very different to insiders than they do to the visitor. When we attribute a Brazilian's tardiness to irresponsibility, or a Moroccan's shifting of attention to their lack of focus, we are being both careless and ethnocentrically narrow-minded. These misinterpretations are examples of what social psychologists call the fundamental attribution error—that, when explaining the behaviors of *others*, there is a pervasive tendency for people to underestimate the influence of the situation and to overestimate others' internal personality dispositions. For example, when I hear strangers lose their temper, I infer that they must be angry people. When I lose my own temper, I blame it on the situation—perhaps the other person was being annoying or the situation was frustrating. After all, I know that I rarely lose my temper, so there must be something unique to this situation that set me off.

An important ingredient in the fundamental attribution error is how much information you have about the person you are judging. The less familiar you are with others, the more likely you are to resort to explanations that reside inside the other person. When we enter foreign environments—which are, by definition, alien—the fundamental attribution error is an accident waiting to happen.

The careful observer would be wise to attend to the advice of Clifford Geertz: "cultural analysis is (or should be) guessing at meanings, assessing the guesses, and drawing explanatory conclusions from the better guesses." Without fully understanding a cultural context, we are likely to misinterpret its people's motives. The result, inevitably, is conflict.

THE TEMPORAL ESTUARY

Being able to shift into the consciousness of another pace of life, no matter which the direction, has its rewards. When event time people learn to accelerate to a clock-driven pace of life, they open doors to otherwise unobtainable wealth and achievements. And when clock time people adapt to slower cultures—well, what could be so painful about entering a consciousness where personal relationships come before accomplishment, in which events are allowed to take their own spontaneous course, where one gives time to time? Taking control of time—learning to live "inside the time"—is an empowering experience. Mastering the temporal concepts of alien cultures becomes its own reward.

Neil Altman, the New York psychoanalyst who once served as a Peace Corps volunteer in southern India, is a poignant example of this. Altman describes the slowness of life, a feeling of time standing still, that he experienced upon setting foot in India. "Leaving the airplane at the Calcutta airport, I entered a small building, the terminal, in fact, where no one was moving. There were several maintenance men carrying small brooms, standing there, with big black eyes, staring at us as we got off the plane and approached the terminal. Ceiling fans rotated languidly in the humid air. There was a sense of having stepped outside time. It just seemed like time was standing still. Not only did nothing seem to be changing from moment to moment, but also there was a sense of continuity with other times produced in me by the slower pace and the absence of machines. It was a feeling of time in which you aren't going anywhere."

For a good year, Altman recalls, he found the stillness of the pace of life very difficult:

At first, it was very stressful, because you're in an unfamiliar situation that makes you feel insecure. It took a year for me to shed my American, culturally based feeling that I had to make something happen. The first year in India, I joined the ranks of "mad dogs and Englishmen" who are the only ones outdoors in the midday sun, as I rode my bicycle in determined search of work to do while everyone else slept. Being an American, and a relatively obsessional American, my first strategy was to find security through getting something done, through feeling worthwhile accomplishing something. My time was something that had to be filled up with progress toward that goal. But it's very anxiety-provoking to encounter people who are in a different sort of time than you are. You say, for example, "So when can I meet you in your field, to talk about planting vegetables?" And they say, "Four o'clock." And then you go there at four o'clock and they aren't there, because they didn't take the appointment literally. And then you get very anxious, because you're trying to get something done and they're not cooperating.

After his initial year, however, Altman not only capitulated to his hosts' sense of time, but began to enjoy it:

By the second year, I relaxed and fully caught on to how one has to live in an Indian village. Since there were no telephones, I would often get up in the morning and ride my bicycle five miles, say, to meet a particular farmer. Arriving there, it was usual to find out that he was away or expected back soon, which might well mean the next day. By the second year, this type of event had stopped feeling like a disappointment, because I no longer really expected to accomplish anything in the first place. In fact, it seemed almost humorous to think that you could truly accomplish what you had set out to accomplish. Instead, I would just go and sit in the local tea shop and meet some new people or simply stare at the animals, children, and other assorted passersby. Then, maybe, something else other than what I'd planned to do would happen. Or maybe it wouldn't. Whatever work was going to get done would come to me. By the second year Indian time had gotten inside me.

Ironically, Altman has found that the domain in his current life most influenced by his Indian temporal experience is his work as a psychoanalyst.

I have used my internalized Indian self often as a psychoanalyst. In my experience, the psychoanalytic session has a culture of its own that is reminiscent of that of an Indian village. One has to enter a session with an openness to the unexpected. The expectations of patient and analyst, the attachment to any particular outcome, get drawn into the process of the session. What is important is to be able to go with the flow of the session, to "be here now," undistracted, as far as possible, by the desire to have something happen other than what is happening. I believe this is something like what the psychoanalyst Wilfred Bion, who interestingly was born in India, meant when he advocated entering each session without memory and desire. I believe that having preset goals in therapy is as futile as planning to get something done in India. If you are going to retain your sanity as a therapist, you need to look with a sense of humor at the pretension that you are going to change the person according to a set plan and schedule. In these ways, living with Indian time has made me a better therapist.[20]

Altman's story eloquently addresses the benefits of a bicultural time sense. Before leaving this chapter, though, it should be noted that achieving literacy in another culture's time system can also lead to a consciousness that goes beyond bitemporal flexibility. Vicente Lopez, when describing his former Chicano commuter culture, argues that his fellow time travelers have done more than master the two cultures' times; they have developed their very own time sense, one unique to their own subculture—what Lopez calls an "estuary" culture. In nature, an estuary is the wide mouth of a river into which the tides flow, an area where the fresh water of the river and the salt water of the sea mix together. "In an estuary," Lopez observes, "nature creates a set of organisms which are not from one side or the other, but completely different. In the same way, people who live on the Tijuana border have this kind of estuarian time. It's not a Mexican time. It's not an American time. It's a different time. The Chicanos are not Americans and are not Mexicans. They live by their own set of rules and have their own unique values and time and pace of life."

And how could it be otherwise? As Oswald Spengler once wrote, "It is by the meaning that it intuitively attaches to time that one culture is differentiated from another."[21] When a new culture is born, so too is a singular time sense.

TEN

MINDING YOUR TIME, TIMING YOUR MIND

> Hold fast the time! Guard it, watch over it, every
> hour, every minute! Unregarded it slips away, like a
> lizard, smooth, slippery, faithless . . . Hold every mo-
> ment sacred. Give each clarity and meaning, each
> the weight of thine awareness, each its true and due
> fulfillment.
>
> THOMAS MANN, *Lotte in Weimar: The Beloved Returns*

There is nothing like studying other cultures to inspire questions about one's own.[1] In a curious way, the outsider's van-
tage point leads us to see home with fresh objectivity and
insight, in the tradition of a de Tocquevillian foreigner. (Then
again, the most profound comment that de Tocqueville had to offer about the time sense of Americans was that they are "always in
a hurry.") More times than not, as the cross-cultural psychologist
Craig Storti observes, "the average expatriate, even the average
tourist, returns from a stay abroad knowing more about his or her
own country than about the one just visited."[2]

For most time travelers, the enduring contribution of cross-
cultural knowledge is what it adds to their lives at home. Once one
understands that there are alternative constructions of time, a new
range of options unfolds. Are there occasions when it would be

healthier to switch to event time? Do I always need to be busy? When might I benefit from doing nothing? In this final chapter, I offer a few suggestions.

ATTEND TO TIME

From my own cultural roots in Judaism I have inherited a philosophy that emphasizes careful attention to time. Judaism is very much a religion of time. It gives more consideration to history and events—the Exodus from Egypt, the revelation of the Torah—than to things. The prophets teach, for example, that the day of the Lord is more sacred than the house of the Lord. Temporal settings, rather than spatial ones, frame the sacred Jewish texts. The Talmud opens with, "From when?" and The Torah with, "In the Beginning."

The observant Jew is guided by a series of time-bound rituals. The day is organized around prayer times. The male infant has his *brit* (ritual circumcision) on his eighth day of life. The *bat* and *bar mitzvah* occur at age thirteen. The traditional year of mourning is highly scheduled: the requirements of the first seven days (*shiva*) are different from those of the first month (*shloshim*), which differ from those of the next eleven months. The Jewish year prescribes eight days for lighting Hanukkah candles, six days for fasting, and eight to eat only unleavened bread. "Counting is our way of noticing," observes writer Letty Pogrebin about her Judaism. "It reminds us that a day counts or it doesn't. Counting imputes meaning; one does not count what one does not value."[3] Even the calendar resists being taken for granted. The modern Jew lives by two sets of dates. I am writing this sentence in the year 1997 on the Gregorian calendar, but when I walk into my local synagogue, where time is measured on the lunar Hebrew calendar, I am in the year 5757.

The essence of Judaism's mindfulness of time is the Sabbath. "And God blessed the seventh day and made it holy," it is written in the Book of Genesis. God devoted six days to creating the heaven and earth. Then, on the seventh, the work is finished—not by building a holy place, but by creating a sacred time. Although the world was created in the first six days, its survival depends upon the holiness of the seventh. "What was created on the seventh day? *Tranquillity, serenity, peace* and *repose.*" The Judaic philosopher Abraham Herschel, who wrote so eloquently about the Sabbath, observed: "Six days a week we seek to dominate the world, on the seventh we seek to dominate the self . . . In the tempestuous ocean of time and toil there are islands of stillness where man may enter a harbor and reclaim his dignity. The island is the seventh day, the Sabbath, a day of detachment from things, instruments and practical affairs as well as of attachment to the spirit."

The Sabbath is not an interlude, but the pinnacle of life. The seventh day, according to Jewish tradition, is a palace in time. It is a sanctuary we build—a temporal sanctuary. Herschel refers to the Sabbath as God's gift of time. ("I have given thee something that belongs to Me. What is that something? A day.") It is pure time; the day when one lives inside time. The Sabbath is our distraction-free opportunity to become masters of time. "Labor is a craft," Herschel reflected, "but perfect rest is an art." And "time," he wrote, "is the presence of God in the world."[4]

The Sabbath ritual also extrapolates to longer time periods. The Torah speaks of every seventh year as a Sabbatical year. In biblical times, the Sabbatical meant the cessation of all agricultural activities (as well as, notably, the cancellation of all outstanding debts). Today, particularly in my academic world, it refers to a time of psychological rest and rejuvenation. Also, Leviticus ordains that every fiftieth year[5], the end of seven Sabbatical cycles, is the sacred Jubilee year. Historically, the Jubilee decreed the cessation of a wide range of activities: land was supposed to be returned to the original owners or their descendants;[6] all Israelites who had been sold into slavery for debt were to be freed; and, as in any Sabbatical year, the land was to lie idle. It is unfortunate that the Jubilee idea has lost its popularity. Perhaps we should bring it back as an emblem for fiftieth birthdays. We could ordain the Jubilee as a year in which we cease production, pause and reflect on where we

have been and where we are going, and, most important, give ourselves up to the dynamics of time and whatever it may bring. It
sure beats the sound of the "big five-oh."

Judaism is not, of course, the only tradition that values time.
One of my favorite temporal cultures is that of the Quiché Indians, who occupy highland villages in Guatemala. The Quiché are
descendants of the Maya, from whom they inherited a great horological tradition. The Mayan calendar was one of the most advanced in the world at the time of the Spanish conquest of Latin
America. In many cases, in fact, the Mayans were more accurate
timekeepers than their European conquerors. But the Mayans, unlike the Europeans, were less concerned with the quantities of
time than with its qualities—in particular, what it meant for human lives.

A particularly interesting aspect of Quiché time is the care with
which they tend to the uniqueness of each day. The day has not
only a proper name, but a divine one. When the Quiché address a
day directly, which they do often, the name is prefixed with the re-
spectful title "*ajaw*"—the equivalent of saying "Greetings Sir, Lord
Thursday." The Quiché believe that every day has its own "face," a
nature, a character, that directs the course of events for each person differently. This nature can be understood by interpreting the
calendar. But a proper reading requires considerable expertise.
The Quiché live with two calendars—a 365-day civil calendar and
a 260-day religious calendar. The latter is in the shape of a wheel,
with neither a beginning nor an end. Each day on the wheel has a
name, a character, and a number, all of which change in different
contexts. To make accurate predictions from this elaborate system, the Quiché rely on specialists: diviners that are known as
"*ajk'ij*," meaning "day keeper." Day keeping is regarded as a sacred
function among the Quiché. The ajk'ij are respected as both
priests and shamans. They communicate directly with the gods to
help lay people decide how to approach each day.

There is an aspect of this day keeping that sounds disturbingly
like the unscientific predictions of astrology. Whether or not this
is true and how much value there is to this criticism is arguable.

But the more crucial lesson from the Quiché is that they think deeply and carefully about each day. There is no such thing as just another Monday. For typical Anglo-Europeans, the temporal challenge of most days is to make every moment productive. The task for the Quiché is more delicate—to decipher how each moment is meant to be lived.[7]

"The main thrust of Torah," according to Jerusalem professor Debbie Weissman, "is to teach us how to spend our limited time on Earth wisely."[8] The same can be said for the temporal philosophy of the Quiché.

LIVING IN MIDDLE TIME

Travel, be it as a tourist or as a cross-cultural psychologist, teaches the aesthetics of middle time: finding a balance between fast and slow, event and clock time, up and down time. When time-urgent workaholics visit relaxed cultures they are pretty sure bets to return home resolved to slow down the regimens of their clock-driven lives. Writer Eva Hoffman captured this striving for moderation when describing her adaption from a childhood in Poland to adolescence in Canada and adulthood in the United States:

> Psychological pleasure is, I think, channeled in time, as physical pain or satisfaction runs along the conduits of our nerves. When time compresses and shortens, it strangles pleasure; when it diffuses into aimlessness, the self thins out into affectless torpor. Pleasure exists in middle time, in time that is neither too accelerated or too slowed down.[9]

Experiments by social psychologists Jonathan Freedman and Donald Edwards confirm Hoffman's observation. Freedman and Edwards found that the relationship between pleasure and time pressure falls into an inverted "U." The greatest pleasure is experienced under an intermediate level of pressure. Too much temporal burden may be stressful, and too little leads to boredom. They also found an inverted U-shaped relationship between time pressure and how well people perform. Once again, the best work occurs at an intermediate degree of time pressure.[10]

Psychologist Mihaly Csikszentmihalyi has found that, at least for Americans, the least happy people are those with no time pressure at all:

> For people in our studies who live by themselves and do not attend church, Sunday mornings are the lowest part of the week, because with no demands on attention, they are unable to decide what to do. The rest of the week psychic energy is directed by external routines: work, shopping, favorite TV shows, and so on. But what is one to do Sunday morning after breakfast, after having browsed through the papers? For many, the lack of structure of those hours is devastating.[11]

It is in the middle ground between too much and too little pressure that people enter the experience, described in an earlier chapter, called "flow." When Csikszentmihalyi kept tabs on people by having them wear beepers, asking them at frequent intervals what they were engaged in and how good they felt, the most positive reports came when people were in moderately challenging activities that engaged their skills. People doing too many things at once tended to be overstressed. But those who were doing nothing at all experienced little sense of flow and little pleasure. Many contemporary psychologists believe the flow experience is an important key to a happy and satisfying life. Studies have shown that flow experiences are not only exhilarating but empowering: they raise self-esteem, competence, and one's overall sense of well-being.[12]

My own studies, we have seen, point to the mixed consequences of a rapid pace of life. People in faster environments are more prone to potentially deleterious stress, as evidenced by higher rates of coronary heart disease; but they are also more likely to achieve a comfortable standard of living and, at least in part because of this, are more satisfied with their lives as a whole. The work by Freedman and Edwards and by Csikszentmihalyi addresses a different domain of psychological satisfaction. They are writing about the pleasure and stress that people find in the very work they are performing as they perform it, while our studies are concerned with how people feel about their lives as wholes. But it is noteworthy that their results, like ours, indicate a mixed blessing from time pressure. Working against the clock is not always stressful, nor is the absence of time pressure inherently relaxing. Time pressure can be energizing and invigorating when served in the right dosage.

Adding to the mixed-blessing argument are the results of a recent study of parents who report that they are under constant time pressure at work. Ellen Greenberger and her colleagues at the University of California at Irvine observed the parenting behavior of 188 employed mothers and fathers of five- to seven-year old children. They found vast differences in how time-pressured parents react to their demands at home. Many follow the stereotypical pattern of returning home exhausted and proceeding to treat their loved ones harshly. But parents with demanding jobs that are also complex, challenging, and stimulating are actually *more* likely to be warm, responsive, and flexible at home.[13] These findings confirm what is also coming out of our research: achieving a balance between the pace of your worklife and the rest of your life may be more important for psychological and physical health than simply whether you work in a high-pressure or low-pressure job.[14]

There is, of course, no single formula for middle time. The difficult task—one of the fine arts of living, really—is to discover the optimal degree of pressure for each person and activity. One strategy I have found personally useful for keeping within the boundaries of middle time is to establish temporal governors: feedback alarms that signal when I am outside of my optimal speed range. Sometimes these signals pop out of unlikely sources. In my own particular case, one of the clearest and most reliable governors has turned out to be an old unpleasant speech impediment, my tendency to stutter when speaking too quickly. I'm certainly not recommending that anyone take up stuttering as a new skill. (Yikes, no!) But my disorder has presented this single psychological gem: more often than not, it prevents my speech cadence from escalating to manic extremes, and it goes off at just about the cutoff point beyond which my speed would otherwise feel unpleasant and unproductive. When younger, I learned to slow down my thoughts to prevent my stuttering. Now, I often use my stuttering to help slow down my thoughts. To my good fortune, my verbal governor has turned out to be an exquisitely tuned gauge. Its permissible speed zone very nicely matches my preferred pace of inner activity.

The British editor and writer Robert McCrum tells of an even more tyrannical temporal governor that recently forced itself into his life. In the summer of 1995, with no warning, he suffered a stroke that has left him seriously physically impaired. In an essay entitled "My Old and New Lives," he described his initial frustration with the slowing it caused in his life:

> In the past, I was noted for the impressionistic speed with which I could accomplish things. At first, the contrast was a source of great frustration. I had to learn to be patient. In English, the adjectival and nominal meanings of "patient" come from the Latin for "suffering" or "endurance"—*patientia*. A patient is by definition "long-suffering."

When preparing McCrum for the post-crisis stage of recovery, one of his doctors warned him how fast the world was going to feel in his new restricted body, offering the prognosis: "You are about to go through the rapids." But one year after the stroke, McCrum had come to appreciate his temporal life change. "I have," he declared, "become friends with slowness, both as a concept and as a way of life."[15]

As Thoreau observed, we must listen for our own drummers. What spells boredom for one person may mean overstimulation to the next. We often hear urbanites talk about "getting out of the rat race" as if it were a sentiment shared by all city dwellers, much as small-town adolescents assume that any sane person would want to "leave the sticks" for more excitement. But the pace of an environment affects people differently. Pace-of-life personality scales have shown that people with both fast and slow temperaments can be found in both fast and slow surroundings. This leads to the most critical art of all—achieving a healthy fit between yourself and your environment.

THE PERSON-ENVIRONMENT FIT

The person-environment fit is related to how much pleasure people experience in everything from their leisure, social, and work lives to the cities and countries where they live. If you tend toward a high activity level and prefer speedy environments, you are best advised to seek out time-pressured jobs and faster places; if your preferences are for a slower pace of life, frenetic jobs and cities can be the (literally, perhaps) kiss of death.

A classic study by Robert Caplan, John French, and their colleagues at the Institute for Social Research at the University of Michigan demonstrates some of the broad consequences of the person-environment fit. Caplan and his colleagues measured the degree of stress and strain experienced by over 2,000 men holding a heterogeneous mix of 23 different white- and blue-collar occupations. By far the best predictor of work stress turned out to be the fit between workers' personal temperaments and the characteristics of their jobs. The degree of fit was more closely connected to psychological stress than were either people's occupations or their personality temperaments considered alone. In other words, the fit is more important than the objective stress level of the job. In fact, after taking into account the person-environment fit, there were virtually no differences in the amount of stress workers experienced across the 23 different occupations. The P-E fit was strongly related to job satisfaction, workload satisfaction, boredom, depression, anxiety, and general irritability. These findings are especially noteworthy because they suggest very applicable directions for increasing job satisfaction: employers often have little control over the characteristics of particular jobs but they can decide who does what task based on workers' temperaments.[16]

The connection between the pace of life and coronary heart disease can likewise be understood as a consequence of the P-E fit. Because fast-paced places and fast-paced people both have higher coronary death rates, we might expect the worst for Type A people living in Type A cultures. But it is overgeneralizing to assume that all individuals follow the pattern of their cultures—a stereotype that cross-cultural psychologists refer to as the ecological fallacy. Type A environments affect people differently. Even Ray Rosenman, one of the co-originators and continuing champions of the Type A concept, has argued that both the person and the environment must be considered when making predictions about coronary heart disease. He argues that distress will be highest for Type A individuals in Type B environments and for Type B individuals in Type A environments.[17]

In the University of Michigan study, for example, the occupation scoring highest on the Type A personality scale was academic administrators. In most universities, these administrators are drawn from the ranks of professors. The work life of academic administrators is typically fast-paced and deadline-driven, while professors tend to have more control over their time. But the Michigan study found that administrators are not necessarily more stressed by their jobs than are their colleagues who remain in the professorial ranks. This may be explained by the fact that the professors attracted to Type A administrative jobs are generally those with Type A temperaments; and these are the very people who are likely to thrive on the temporal demands of these jobs. In the same way, the New Yorker addicted to the buzz of Wall Street may be best off staying right where he or she is.

Coronary heart disease remains the single largest cause of death in the United States. Traditional (non-Type A) risk factors, such as diet, smoking, and essential hypertension, can account for only 50 percent of its causes.[18] If, as it appears, the temporal matching of persons to environments comprises a significant portion of the remaining 50 percent, it seems worth the effort to make these fits work.

One provocative paradigm for understanding the person-environment fit comes from the research on social entrainment being conducted by psychologist Joseph McGrath and his colleagues at the University of Illinois. The entrainment concept, borrowed from biology, refers to the process by which one temporal rhythm is captured and modified by another. For example, a flock of birds becomes a flock through entrainment: each member using delicately tuned sensory machinery to detect and then adjust to environmental cues. Similarly, herding animals become a herd by synchronizing their rhythms to each other. On human byways, motor vehicle drivers manage to travel at high speeds for long distances (mostly) without running into one another by mutual entrainment. McGrath and his colleagues have demonstrated in their experiments that entrainment surfaces in a wide range of situations, ranging from how quickly people work to the tempo of their social encounters.[19]

The possibility of entrainment raises a number of questions. Just how far can people adapt themselves to temporally disagreeable environments and vice versa? Under what conditions is entrainment most likely to occur, and when is it most successful? What sort of person is most suited to temporal adaptation? Can we design environments that are capable of adapting themselves to the preferred rhythms of individuals?

The possibility of entrainment underscores the importance of calibrating our own rhythms to those around us. If we aspire toward what Carl Jung called synchronicity, there must be temporal flexibility on both sides of the person-environment equation.

MULTITEMPORALITY

There is an interesting concept in personality psychology known as psychological androgyny. Traditional research on psychological sex differences has treated masculinity and femininity as polar opposites. Early studies offered the unsurprising finding that stereotypical "masculine"-type personalities are more likely to succeed in situations calling for qualities like assertiveness, while "feminine" types do better in situations requiring such traits as nurturance and emotional expression. Since professional success has traditionally been associated with masculine skills, one result of this stereotype is that a long line of career women have had to confront doubts about their femininity; the more successful they were, the more they felt they were looked upon as more like a man than a woman.

But a recent generation of (female) social psychologists—leaders like Sandra Bem and Janet Spence—have challenged the assumption that women's ability to compete in traditionally masculine endeavors must mean a compromise of their feminine side. It is they who have developed the idea of psychological androgyny. Psychologically androgynous people combine both traditional masculine traits (such as assertiveness) and feminine traits (such as nurturance) in their personality repertoire. The androgynous person is not a psychological neuter who falls midway between extreme masculinity and femininity, but one who has both strong masculine and feminine attributes at his or her disposal.

A number of studies have demonstrated the value of psychological androgyny. Whereas masculine types do better in traditional "male" situations and feminine types excel in "female" situations, experiments have shown that androgynous people—both men and women—are more likely to succeed at *both* masculine and feminine tasks. It has been demonstrated, for example, that masculine and androgynous personalities are better than feminine types at resisting group pressure to conform, but that feminine and androgynous people do better on tasks such as counseling a fellow student with problems.[20] Androgynous and feminine spouses—both husbands and wives—also tend to have happier marriages.[21] The androgynous person, in other words, has access to the best of both worlds.

The pace of life moves in an analogous pattern. "The question is not just what floor of the building you're living on," as the transpersonal psychologist Ken Wilber puts it, "but how many floors you have access to as you negotiate your way through life."[22] Many situations are best met by a temporal approach requiring a rapid pace of life: speed, attention to the clock, a future orientation, the ability to value time as money. Other domains in life—rest, leisure, the incubation of ideas, social relationships—are more adequately met with a relaxed attitude toward time. The person, or the culture, who combines both modes in a temporal repertoire—or even better, who can draw upon a multiplicity of modes—is more likely to be up to all occasions. Jeremy Rifkin speaks of the dangers of temporal ghettos. People who are confined to rigid and narrow temporal bands are unprepared to determine their own futures and political fates.[23] Multitemporality is the ticket out of these temporal ghettos. To have the ability to move quickly when the occasion demands it, to let go when the pressure stops, and to understand the many temporal shades of grey may be the real answer to the question of "Which pace of life is best?" As Lewis Mumford wrote:

> Though our first reaction to the external pressure of time necessarily takes the form of the slow-down, the eventual effects of liberation will be to find the right tempo and measure for every human activity; in short, to keep time in life as we do in music, not by obeying the mechanical beat of the metronome—a device only for beginners—but by finding the appropriate tempos from passage to passage, modulating the pace according to human need and purpose.[24]

Like the psychological androgyne, the truly multitemporal person and culture does not simply fall in the average range, but has the ability to move as rapidly or slowly as is needed. Not surprisingly, another significant result of the University of Michigan study was that personal flexibility (versus rigidity) was an effective buffer against stress and job dissatisfaction no matter what one's occupation. The European who works hard enough to achieve, but who can decelerate to enjoy *la dolce vita*, the fruit of his or her labors, possesses an element of this multitemporality. The Japanese worker who excels at both speed and slowness also understands the skill. This is hardly to say that all European and Japanese workers have mastered multitemporality. If anything, in fact, the data highlight the price that many in these countries are paying for their rapid pace of life; it is no coincidence that coronary heart disease rates in Western Europe are some of the highest in the world and that suicide is a serious problem in Japan. But the traditional values in these cultures offer potential templates—recipes of a sort—that illuminate paths by which the mindful individual may take control of his or her time.

Joyce Carol Oates wrote, "Time is the element in which we exist . . . We are either borne along by it or drowned in it." How to be productive enough to be comfortable, to minimize the temporal stress on which this achievement is built, and to simultaneously make time for caring relationships and a civilized society—this is the multitemporal challenge.

SEIZE THE CONTROLS

And the end of all exploring will be to arrive where we started
And know the place for the first time
                                T. S. ELIOT, *The Four Quartets*

One last story. While planning what was to turn into my twelve-month trip around the world, it seemed as if every seasoned and/or frustrated traveler I met offered words of advice. These ranged from lists of places that I absolutely *had* to visit or avoid, to graphic descriptions of what would happen to my body if I even thought about drinking the water. But the single most prophetic wisdom came from an unlikely source. While sitting in the chair of a rather unworldly dentist, my mouth stuffed with the usual unpleasant objects, he offered the longest nondental communication that ever passed between us: "I went to another country once. You learn a lot about yourself."

He was right on the money. After a year of vagabonding across some twenty countries on three continents, visiting every marvel in touring distance from the Great Wall of China to the Wailing Wall in Jerusalem, having collected multinational data on what would turn out to be the focus of my professional research ever since, what I mostly carried home with me was a new point of view.

My most lasting insights, those that continue to make a difference in how I live my life, always seem to hover about the theme of time. When people live on the road for extended periods, there seems to be a point when they shift into the consciousness of a transient. Most travelers I have questioned about this transition report that the critical cutoff seems to occur sometime around three months. After that, the days of the week and even the months of the year—especially for those who have the good sense to follow the warm weather—meld into one another. Expectations and plans for the future become stunted, or nonentities.

There is something about the frequent, swift, and often dramatic changes that are the fabric of long-term traveling—deciding in the middle of breakfast to pack up and head for another country before check-out time; ending a seemingly intimate affair because one partner is inspired to head east while the other selects west—that usually leaves no alternative other than to live from day to day. The force is so strong that it feels more somatic than a volitional choice. I know that personally, by the end of each extended excursion, it felt as if I was physically incapable of fixing my thoughts on either the future or the past. This is not to presume that I had transcended time to some idyllic Zen present-connectedness; more often than not I was, as we say in psychology, simply out to lunch. It was temporal limbo. Having spent most of my life until then as a future-oriented person—a future often defined by the expectations of others—I found it almost comical to observe how my mind was unable to focus on what was coming tomorrow, even when it might be an event that I had been anticipating for months, such as my first visit to the Great Pyramids. But it was definitely the beat of my own clock. As the philosopher Johann Herder once wrote, "everything transient has the measure of *its* time within itself."

So when I arrived home to resume my role as a university professor it was with the sensibility of a vagabond. My intellect was temporally unavailable, and I was disoriented from many of the usual ingredients of culture shock. Many long-term travelers will tell you that the shock of returning home is often more jarring than that of leaving. I believe this is because we return with the dangerous illusion that, having arrived home, it is at last permissible to let up, to cease the hard work of coping with constant change. (It is for good reason that the word "travel" is related to the French "*travail*," meaning hard work and penance.) But social psychologists will tell you that it is at those very moments when people hold an "illusion of invulnerability" that they are really the most susceptible targets.

I walked up to my university office trying my best to look like Mr. Chips, but feeling more like Rip Van Winkle. I was frightened, afraid of trying to go home again—not in the sense that Thomas Wolfe meant, but I feared that I would not be able to reenact my expected professional role. What happened next took me by surprise. Suddenly, as if a toggle had been switched, I felt my mind morphing, like a cheap movie image, into the very time-driven psyche that had stood on this same spot twelve months earlier. It seemed as if every task expected of "Professor Levine" came rushing back; I knew what I had to do and I knew the time and place to do it. For a full year my university had gotten along just fine, thank you, without me. And now, with frightening immediacy, my future was once again filled with an abundant helping of "shoulds" and "musts." My schedule was packed.

Part of me welcomed the structure and normalcy. But a louder voice protested that coming home needn't mean tossing out the proverbial baby with the bathwater. Must I repress all of the changes I had gone through? In one of those rare moments of inner sanity, the louder voice ordered me to leave the building in order to think things through.

When I was able to consider my situation from a detached perch, it occurred to me that my culture shock was simply a limbo state, a bridge between two modes of life: the day-to-day spontaneity of longterm travel on the one hand and the schedule-driven existence of my professional life on the other. In fact, I was sandwiched between the very temporal forces I had been studying. Perhaps it was simply some undigested Nepalese curry, but I found myself thinking of the Tibetan Book of the Dead, the sacred Buddhist text that teaches how to consciously control the processes of dying, death, and rebirth. Death, the book instructs, is an intermediate state between life and one's inevitable (with the exception of a few super Buddha-types) reincarnation on earth. The after-death transition must be handled with care and precision, as it offers the singular opportunity for consciously shaping the character of one's next life:

> *Improvident art thou in dissipating thy great opportunity; mistaken, indeed, will thy purpose be now if thou returnest empty-handed.*
>
> The Tibetan Book of the Dead[25]

It felt as if I, too, was passing between incarnations. My charge now was to discover what I could retain from the life I was leaving to enhance the one I was approaching. And at that moment I was offered a rather simple insight that improved my life: I understood I had a rare opportunity to break cycles of worthless habits and compulsions. I resolved that each time I saw myself reentering a pretrip activity—be it a professional task, such as meeting with a student, teaching a class or writing up a research paper; or a social activity, anything from going to lunch with a colleague to exchanging niceties with an acquaintance to answering the telephone—I would be alert to intercept my knee-jerk response. And I would pause; then I would ask two questions. First, is this something that I absolutely *must* do? And, second, is it something that I *choose* to do? Unless there was a "yes" to one of these questions I would not invest my time in the endeavor.

Over the next couple of weeks I posed my questions compulsively. I discovered that the answer to the "must" question mostly came back "no"; certainly, there were many more negatives than I had anticipated. I recognize, of course, that my potential "no" quotient may be higher than that of most people. After all, I am blessed with a profession that allows a formidable degree of personal control. I was also, at the time, without the responsibilities of either a marriage or children. But still, it was surprising how painlessly my colleagues and companions usually seemed able to survive without my participation. This modal "no" answer mostly left me evaluating my options by the second criterion. And here I was surprised at the number of my "yes" responses. Most unexpected was how often I chose to engage in relatively mundane activities.

When all was said and done, when my transition period had passed, I felt more in control of my life than I ever had before, and it is a feeling that has stayed with me to this day. I understand that my time truly is *my* time. And even though the pace of my life, like everyone else's, is often directed by the world around me, it has become clear that people have considerably more control over their tempos than they often let themselves believe. And I've come to see another basic truth: that our time is our life. As Miles Davis said, "Time isn't the main thing. It's the only thing."[26] How we construct and use our time, in the end, defines the texture and quality of our existence. To seize control over the structure of one's time is my own definition of what it means, as it is said in the Tibetan Book of the Dead, to avoid "devoting thyself to the useless doings of this life."

And that, more than anything, is what I have taken away from my studies of the time senses of other cultures. Borrowing again from Russell Banks's image of Hawthorne's Wakefield: I had moved out of my house and this is what appeared when I looked back to "see what's true there."[27] Simply that. It may not hold a candle to the "primary clear light" that the Tibetan Buddhists watch for at the entrance to the afterlife. But I do know that my time has been just a little more my own ever since.

NOTES

PREFACE
TIME TALKS, WITH AN ACCENT

1. Hall, Edward T. (1959). *The Silent Language.* Garden City, N.Y.: Doubleday.
2. Strauss, A. L. (1976). *Images of the American City.* New Brunswick, N.J.: Transaction Books.
3. *Time* Magazine, March 11, 1985.
4. For a good example, see: Keyes, R. (1991). *Timelock.* New York: HarperCollins.
5. The New York Times Book Review (1991, August 18). Itchy feet and pencils: A symposium. *The New York Times Book Review*, 3, 23.

CHAPTER ONE
TEMPO: THE SPEED OF LIFE

1. Spradley, J. P., and Phillips, M. (1972). Culture and stress: A quantitative analysis. *American Anthropologist* 74, 518–29.
2. Personal communication, January 16, 1996.
3. In most countries, data were collected in either the largest city or a rival major city: Amsterdam (The Netherlands), Athens

(Greece), Budapest (Hungary), Dublin (Ireland), Frankfurt (Germany), Guanzhou (China), Hong Kong (Hong Kong).

4. Jakarta (Indonesia), London (England), Mexico City (Mexico), Nairobi (Kenya), New York City (U.S.A.), Paris (France), Rio de Janeiro (Brazil), Rome (Italy), San Jose (Costa Rica), San Salvador (El Salvador), Seoul (South Korea), Singapore (Singapore), Stockholm (Sweden), Taipei (Taiwan), Tokyo (Japan), Toronto (Canada), and Vienna (Austria). In four other countries, for various reasons, the observations were made in more than one city. In Poland, data were collected in Wroclaw, Lodz, Poznan, Lublin, and Warsaw. In Switzerland, measures were taken in both Bern and Zurich. In Syria and Jordan, most observations were made in the capital cities of Damascus and Amman, but some were done in secondary population centers. In each of these cases, data from the different cities were combined for that country. Data for all countries were collected during the summer or other warm-weather months of the year, over the period 1992–95.

5. Hoch, I. (1976). City size effects, trends and policies. *Science* 193, 856–63, 857. For further discussion of the economic hypothesis, see: Bornstein, M. H. (1979). The pace of life: Revisited. *International Journal of Psychology* 14, 83–90.

6. Statistics were based on the most recent available data from the World Bank. World Bank (1994). *The World Bank Atlas: 1995.* Washington, D.C.: World Bank.

7. Henry, J. (1965, March/April). White people's time—colored people's time. *Trans-Action* 2, 31–34.

8. Ibid., p. 24.

9. Horton, J. (1972). Time and cool people. In Kochman, T. (ed.), *Rappin and Stylin' Out*, 19–31. Urbana, Ill.: University of Illinois Press.

10. Hunt, S. (1984, May 25). Why tribal peoples and peasants of the Middle Ages had more free time than we do. *Maine Times*, p. 40.

11. Johnson, A. (1978, September). In search of the affluent society. *Human Nature*, 50–59.

12. Schor, J. B. (1991). *The Overworked American.* New York: Basic Books, 10.

13. Hall, E. T. (1959). *The Silent Language.* New York: Doubleday.

14. Bohannan, P. (1980). Time, rhythm, and pace. *Science 80*, 1, 18–20.

15. Niles, S. E-mail posting on *Intercultural Network*, May 19, 1995.

16. Personal communication, November 21, 1993.

17. Since our 31 countries study was based chiefly on the largest

city in each country, it did not provide a particularly strong retest of the population size hypothesis.

18. Wright, H. F. (1961). The city-town project: A study of children in communities differing in size. Unpublished grant report.

19. Amato, P. R. (1983). The effects of urbanization on interpersonal behavior. *Journal of Cross-Cultural Psychology* 14, 353–67.

20. Bornstein, M. H. (1979). The pace of life: Revisited. *International Journal of Psychology* 14, 83–90.

21. Bornstein, M., and Bornstein, H. (1976). The pace of life. *Nature* 259, pp. 557–59.

22. Some evidence for this hypothesis may also be found in a study by: Hoel, L. A. (1968). Pedestrian travel rates in central business districts. *Traffic Engineer* 38, 10–13.

23. More sophisticated temperature-humidity indices are not readily available for many international cities.

24. Triandis, H. (1994). *Culture and Social Behavior.* New York: McGraw-Hill.

25. Individualism-collectivism scores were provided by Harry Triandis.

26. Bourdieu, Pierre (1963). The attitude of the Algerian peasant toward time. In: Pitt-Rivers, J. (ed.) *Mediterranean Countrymen,* 55–72. Paris: Mouton.

27. Friedman, M., and Rosenman, R. H. (1959). Association of specific overt behavior patterns with blood and cardiovascular findings. *Journal of the American Medical Association* 240, 761–63.

28. Jenkins, C. D., Zyzanski, S. J., and Rosenman, R. H. (1979). *Jenkins Activity Survey: Form C.* New York: Psychological Corporation.

29. Wright, L., McCurdy, S., and Rogoll, G. (1992). The TUPA Scale: A self-report measure for the Type A subcomponent of time urgency and perpetual activation. *Psychological Assessment* 4, 352–56.

30. Keyes, R. (1991). *Timelock.* New York: HarperCollins.

31. For more detailed self-evaluation measures of time urgency, two good sources are: Landy, F. J., Restegary, H., Thayer, J., and Colvin, C. (1991). Time urgency: The construct and its meaning. *Journal of Applied Psychology* 76, 644–57; or: Friedman, M., Fleischmann, N., and Price, V. (1996). Diagnosis of Type A behavior pattern. In Robert Allan and Stephen Scheidt (eds.), *Heart and Mind: The Practice of Cardiac Psychology,* 179–96. Washington, D.C.: American Psychological Association.

32. Ulmer, D. K., and Schwartzburd, L. (1996). Treatment of time pathologies. In Robert Allan and Stephen Scheidt (eds.), *Heart and*

Mind: The Practice of Cardiac Psychology, 329–62. Washington, D.C.: American Psychological Association.

33. For further information about the scale, see: Levine, R., and Conover, L. (1992, July). The pace of life scale: Development of a measure of individual differences in the pace of life. Paper presented to the International Society for the Study of Time, Normandy, France; and: Soles, J. R., Eyssell, K., Norenzayan, A., and Levine, R. (1994, April). Personality correlates of the pace of life. Paper presented at the meetings of the Western Psychological Association, Kona, Hawaii.

34. Dapkus, M. (1985). A thematic analysis of the experience of time. *Journal of Personality and Social Psychology* 49, 408–19.

CHAPTER TWO
DURATION: THE PSYCHOLOGICAL CLOCK

1. Reported in: Campbell, S. (1990). Circadian rhythms and human temporal experience. In Block, R. (ed.). *Cognitive Models of Psychological Time*, 101–18. Hillsdale, N.J.: Lawrence Erlbaum.

2. Block, R. A. (1994). Temperature and psychological time. In Macey, S. L. (ed.) *Encyclopedia of Time*, 594–95. New York: Garfield.

3. Block, R. (1990). Models of psychological time. In Block, R. (ed.). *Cognitive Models of Psychological Time*, 1–36. Hillsdale, N.J.: Lawrence Erlbaum.

4. Macleod, R. B., and Roff, M. F. (1936). An experiment in temporal disorientation. *Acta Psychologica* 1, 381–423.

5. See: Campbell, S. (1990). Circadian rhythms and human temporal experience. In Block, R. (ed.) *Cognitive Models of Psychological Time*, 101–18. Hillsdale, N.J.: Lawrence Erlbaum.

6. Aschoff, J. (1985). On the perception of time during prolonged temporal isolation. *Human Neurobiology* 4, 41–52.

7. Campbell, S. (1990). Circadian rhythms and human temporal experience. In Block, R. (ed.) *Cognitive Models of Psychological Time*, 101–18. Hillsdale, N.J.: Lawrence Erlbaum.

8. Aschoff, J. (1985). On the perception of time during prolonged temporal isolation. *Human Neurobiology* 4, 41–52.

9. Siffre, M. (1964). *Beyond Time*, 118, 182. New York: McGraw-Hill.

10. Buckhout, R. (1977). Eyewitness identification and psychology in the courtroom. *Criminal Defense* 4, 5–10.

11. Loftus, E. F., Schooler, J. W., Boone, S. M., and Kline, D.

(1987). Time went by so slowly: Overestimation of event duration by males and females. *Applied Cognitive Psychology* 1, 3–13.

12. Ibid., 3.

13. Schneider, A. L., Griffith, W. R., Sums, D. H., and Burcart, J. M. (1978). *Portland Forward Records Check of Crime Victims*. Washington, D.C.: U.S. Department of Justice.

14. Veach, T. L., and Touhey, J. C. (1971). Personality correlates of accurate time perception. *Perceptual and Motor Skills* 33, 765–66.

15. Gardner, R. M., Brake, S. J., and Salaz, V. E. (1984). Reproduction and discrimination of time in obese subjects. *Personality and Social Psychology Bulletin* 10 (4), 554–63.

16. Andrew, J. M., and Bentley, M. R. (1976). The quick minute: Delinquents, drugs, and time, *Criminal Justice & Behavior* 3 (2), 179–86.

17. Many of these findings are summarized in: Orme, J. E. (1969). *Time, Experience and Behavior*. London: Iliffe Books.

18. Friedman, W. (1990). *About Time: Inventing the Fourth Dimension*. Cambridge, Mass.: MIT Press.

19. Suzuki, D. T. (1959). *Zen and Japanese Culture*. New York: Pantheon Books.

20. Murphy, M., and White, R. (1978). *The Psychic Side of Sports*, 46. Reading, Mass.: Addison-Wesley.

21. Ibid., 45.

22. Zimbardo, P. G., Marshall, G., and Maslach, C. (1971). Liberating behavior from time-bound control: expanding the present through hypnosis. *Journal of Applied Social Psychology* 1, 305–23.

23. Cooper, L. F., and Erickson, M. H. (1959). *Time Distortion in Hypnosis*. Baltimore: Williams and Wilkins.

24. Huxley, A. (1962). *Island*. New York: Harper and Row.

25. There is a large body of psychological research concerning individual differences in the "optimal arousal level." See, for example: Mehrabian, A., and Russell, J. (1974). *An Approach to Environmental Psychology*. Cambridge, Mass.: MIT Press.

26. Melges, F. T. (1982). *Time and the Inner Future*, 177. New York: John Wiley.

27. Ibid., xix.

28. Harton, J. J. (1939). An investigation of the influence of success and failure on the estimation of time. *Journal of General Psychology* 21, 51–62.

29. Ornstein, R. (1977). *The Psychology of Consciousness* (2nd ed.). New York: Harcourt, Brace, Jovanovich.

30. Meade, R. D. (1960). Time on their hands. *Personnel Journal* 39, 130–32.

31. Reported in: Cialdini, R. (1993). *Influence: Science and Practice* (3rd ed.), 197. New York: HarperCollins.

32. Hall, E. T. (1959). *The Silent Language*, 152–53. New York: Doubleday.

33. Friedman, W. (1990). *About Time: Inventing the Fourth Dimension.* Cambridge, Mass.: MIT Press.

34. Cahoon, D., and Edmonds, E. M. (1980). The watched pot still won't boil: Expectancy as a variable in estimating the passage of time. *Bulletin of the Psychonomic Society* 16, 115–16.

35. Weaver, M. A. (1991, October 7). Brunei. *The New Yorker*, 64.

36. Hoffman, E. (1993). *Exit into History: A Journey through the New Eastern Europe*, 78. New York: Penguin.

37. Ibid., 282.

38. Ueda, K. Sixteen ways to avoid saying "no" in Japan. In Condon, J., and Saito, M. (1974). *Intercultural Encounters with Japan.* Tokyo: The Simul Press.

39. Electronic mail messages, April 29-May 2, 1995, and September 7, 1995.

40. Ibid.

41. Callus, H. (1959). *China: Confucian and Communist*, 37. New York: Henry Holt.

42. Keyes, R. (1991). *Timelock.* New York: HarperCollins.

43. Levy, J. (1974). Psychological implications of bilateral asymmetry. In Dimond, S. J., and Beaumont, J. G., *Hemisphere Function of the Human Brain.* New York: John Wiley & Sons.

The findings by Roger Sperry and his colleagues are based on the study of so-called split-brain patients. These are individuals who have been either victims of brain damage or recipients of surgical procedures that resulted in severance of the corpus callosum—the thick nerve cable responsible for communication between the two hemispheres of the brain, which is a characteristic of "normal" consciousness. Without an intact corpus callosum, the researchers found, the two hemispheres of the brain can be induced to work independently. By presenting different tasks to each hemisphere in these split-brain patients, Sperry and his successors have been able to catalog the tasks for which each hemisphere takes primary responsibility. In normal people, however, the corpus callosum provides constant communication between the two hemispheres, so that both hemispheres work together on virtually every task. In other words, although it is helpful to talk in general terms about left- and right-hemisphere modes of think-ing in normal people, it is scientifically unjustified to employ the cur-rent pop-psychology lingo about left- and right-hemisphere thinking in a literal sense—as if one side of the brain were lit up and the other closed down, depending on what was taking place. For more on this topic, see Howard Gardner's chapter on "What We Know (and Do Not Know) About the Two Halves of the Brain" in his book *Art, Mind and Brain* (1982).

44. Edwards, B. (1979). *Drawing on the Right Side of the Brain.* Los Angeles: Houghton-Mifflin.

45. Csikszentmihalyi, M. (1990). *Flow: The Psychology of Optimal Experience.* New York: Harper & Row.

46. Both quotes from: Myers, D. (1992). *The Pursuit of Happiness,* 133. New York: Avon.

47. Csikszentmihalyi, M. (1988). The flow experience and its signif-icance for human psychology. In Csikszentmihalyi, M., and Csikszent-mihalyi, I. *Optimal Experience: Psychological Studies of Flow in Consciousness.* Cambridge: Cambridge University Press.

48. Csikszentmihalyi, M., *Flow: The Psychology of Optimal Experience,* 81.

49. Capra, F. (1975). *The Tao of Physics.* Boulder, Colo.: Shambhala.

50. Rucker, R. (1984). *The Fourth Dimension: A Guided Tour of the Higher Universes.* Boston: Houghton Mifflin.

51. Pinker, S. (1994). *The Language Instinct,* 209. New York: William Morrow; quoted in an e-mail posting by Mark Aultman on ISSTL@PSUVM.PSU.EDU, February 26, 1996.

CHAPTER THREE
A BRIEF HISTORY OF CLOCK TIME

1. Lewis, J. D., and Weigart, A. J. (1981). The structures and meaning of social time. *Social Forces* 60, 432–62.

2. Lightman, A. (1993). *Einstein's Dreams,* 150–51. New York: Pan-theon.

3. There are a number of excellent sources providing a more com-plete description of the history of timepieces. See, for example: Boorstin, D. J. (1983). *The Discoverers.* New York: Random House.

4. Ibid., 33.

5. Kahlert, H., Muhe, R., Brunner, G. L. (1986). *Wristwatches: History of a Century's Development,* 12–13. West Chester, Pa.: Schiffer Publishing.

6. Cited in Kahlert, et al. (1986), 11.

7. Greenhill, J. (1993, April 23). Running late? Never in a million years. *USA Today*, 1.

8. Hawking, S. (1988). *A Brief History of Time: From the Big Bang to Black Holes*. New York: Bantam.

9. The Szalai quote above is Szalai, A. (1966). Differential evaluation of time budgets for comparative purposes. In Merritt, R., and Rokkan, S. (eds.) *Comparing Nations: The Use of Quantitative Data in Cross-National Research*, 239–358. New Haven: Yale University Press.

10. Bloch, M. (1961). *Feudal Society*. Chicago: University of Chicago Press.

11. Aveni, A. (1989). *Empires of Time*. New York: Basic Books.

12. Quoted in: Keyes, R. (1991). *Timelock*, 20. New York: Harper-Collins.

13. Mumford is quoted in: Westergren, G. (1990). *Time: Experiences, Perspectives and Coping-Strategies*, 8. Stockholm, Sweden: Almqvist and Wiksell.

14. Keyes, R. (1991). *Timelock*, 18. New York: HarperCollins.

15. O'Malley, M. (1990). *Keeping Watch: A History of American Time*. New York: Viking.

16. Ibid., 40.

17. *U.S. News & World Report* (1990, October 22). The times of our lives: Conversation with Michael O'Malley, 66.

18. O'Malley, M. *Keeping Watch*, 95.

19. Quoted in ibid., 156.

20. Ibid., 157.

21. Ibid., 161.

22. Waterbury Watch Company (1887). *Keep a Watch on Everybody*. New York: BAC, Bx 6, "Waterbury Watch Co." folder.

23. O'Malley, op. cit., 148.

24. Cawelti, J. G. (1965). *Apostles of the Self-Made Man*, 118. Chicago: University of Chicago Press.

25. Rifkin, J. (1987). *Time Wars*, 110. New York: Henry Holt.

26. Ibid., 111.

27. Braverman, H. (1974). *Labor and Monopoly Capital*, 321. New York: Monthly Review Press.

28. J. Rifkin, *Time Wars*, 111.

29. Ibid., 124.

30. Ibid., 134.

31. Ibid., 136.

32. Ibid., 169.

33. Ibid. 145.

34. Ibid., 15.

35. Haley, A. (1986, March 16). Writer's guide. *Los Angeles Times Magazine*, 16.

36. Rimer, S. (1993, July 13). They measure time by feet. Reprinted in *The Fresno Bee*, A1, A8.

37. Rifkin, J. *Time Wars*, 1.

38. Shaw, J. (1994). Punctuality and the everyday ethics of time. *Time and Society* 3, 79–97, 86, 87.

39. Personal communication. For a report of his work with the metronome, see: Kir-Stimon, William (1977). "Tempo-statis" as a factor in psychotherapy: Individual tempo and life rhythm, temporal territoriality, time planes and communication. *Psychotherapy: Theory, Research and Practice* 14, 245–48.

40. Lauer, R. H. (1981). *Temporal Man: The Meaning and Uses of Social Time*. New York: Praeger.

41. Zerubavel, E. (1977). The French Revolution calendar: A case study in the sociology of time. *American Sociological Review* 42, 870.

42. Zerubavel, E. (1981). *Hidden Rhythms: Schedules and Calendars in Social Life*. Chicago: University of Chicago Press.

43. Rifkin, J. *Time Wars*, 2, 5.

44. Meeker, J. Reflections on a digital watch. Quoted in *Utne Reader* (1987, September/October), 57.

CHAPTER FOUR
LIVING ON EVENT TIME

1. Lauer, R. (1981). *Temporal Man: The Meaning and Uses of Social Time*. Praeger: New York.

2. Jones, J. (1993). An exploration of temporality in human behavior. In: Schank, R., and Langer, E. (eds.) *Beliefs, Reasoning, and Decision-Making: Psycho-Logic in Honor of Bob Abelson*. Hillsdale, N.J.: Lawrence Erlbaum.

3. Schachter, S., and Gross, L. (1968). Manipulated time and eating behavior. *Journal of Personality and Social Psychology* 10, 93–106.

4. Castaneda, J. (1995, July). Ferocious differences. *The Atlantic Monthly*, 68–76, 73, 74.

5. Bock, P. (1964). Social structure and language structure. *Southwestern Journal of Anthropology* 20, 393–403.

6. Lauer, R. H. (1981). *Temporal Man: The Meaning and Uses of Social Time.* New York: Praeger.

7. Sorokin, P. (1964). *Sociocultural Causality, Space, Time.* New York: Russel and Russel.

8. Rifkin, J. (1987, September/October). Time wars: A new dimension shaping our future. *Utne Reader,* 46–57.

9. Thompson, E. P. (1967). Time, work-discipline, and industrial capitalism. *Past and Present* 38, 56–97.

10. Raybeck, D. (1992). The coconut-shell clock: Time and cultural identity. *Time and Society* 1 (3), 323–40.

11. Leach, E. R. (1961). *Rethinking Anthropology.* London: The Athlone Press.

12. Hall, E. (1983). *The Dance of Life.* Garden City, N.Y.: Doubleday.

13. Gonzalez has also done important research on the subject of time. See Gonzalez, A., and Zimbardo, P. (1985, March). Time in perspective. *Psychology Today,* 20–26.

14. Hall, E. (1983). *The Dance of Life.* Garden City, N.Y.: Doubleday.

15. Bluedorn, A., Kaufman, C., and Lane, P. (1992). How many things do you like to do at once? An introduction to monochronic and polychronic time. *Academy of Management Executive* 6, 17–26.

16. UPI (1985, June 23). Ships with 1,800 Marines off Lebanon. Reprinted in *The Fresno Bee,* A1.

CHAPTER FIVE
TIME AND POWER: THE RULES OF THE WAITING GAME

1. Osuna, E. (1985). The psychological cost of waiting. *Journal of Mathematical Psychology* 29, 82–105.

2. Gwertzman, B. (1969, May 13). Soviet shoppers spend years in line. *New York Times,* 13.

3. Dressler, C. (1988, June 21). Minute here, an hour there: They add up. *The Fresno Bee,* A1.

4. Undoubtedly the best source of information on this topic is: Schwartz, B. (1975). *Queuing and Waiting.* Chicago: University of Chicago Press. The present chapter draws heavily on Schwartz's pioneering work.

5. Quoted in: Gibbs, N. (1989, April 24). How America has run out of time. *Time,* 58–67, 67.

6. Robinson, J. (1991, November). Your money or your time. *American Demographics,* 22–25.

7. Mehta, Ved (1987, January 19). Letter from New Delhi. *The New Yorker*, 52–69.

8. Ibid., 58.

9. Cialdini, R. (1988). *Influence: Science and Practice, 2nd Ed*, 230. Glenview, Ill.: Scott-Foresman.

10. Halpern, J., and Isaacs, K. (1980). Waiting and its relation to status. *Psychological Reports* 46, 351–54.

11. See: Levine, R., West, L., and Reis, H. (1980). Perceptions of time and punctuality in the United States and Brazil. *Journal of Personality and Social Psychology* 38 (4), 541–50.

12. From E. B. White (1935). *The Second Tree from the Corner*. New York: Harper and Bros. 225–26.

13. Schwartz, B. (1975). *Queuing and Waiting*. Chicago: University of Chicago Press, 21.

14. Ibid., 110–32.

15. Gibbs, N. (1989, April 24). How America has run out of time. *Time*, 67.

16. Quoted in: Schwartz, B. (1975). *Queuing and Waiting*, 135. Chicago: University of Chicago Press.

17. Ibid., 135–52.

18. Solzhenitsyn, A. (1968). *The Cancer Ward*, 222. New York: Dial.

19. Clifford, C., with Holbrooke, R. (1991, May 6). Annals of Government (The Vietnam Years)—Part 1. *The New Yorker*, 79.

20. Ibrahim, Y. M. (1991, February 3). In the mideast, a fear that war is only the beginning. *The New York Times Week in Review*, 1–2, 1.

21. Greve, F., and Donnelly, J. (1991, January 27). Oil nears water supply. *The Fresno Bee*, A1, A8.

22. Allman, T. D. (1996, June 17). Saddam wins again. *The New Yorker*, 60–65.

23. Gabler, N. (1994). *Winchell: Gossip, Power and the Culture of Celebrity*, xv. New York: Knopf.

24. *New York Times*, November 25, 1963; quoted in: Schwartz, B. (1975). *Queuing and Waiting*, 42. Chicago: University of Chicago Press.

25. Post, E. (1965). *Emily Post's Etiquette: The Blue Book of Social Usage*, 48. New York: Funk and Wagnall's.

26. Quoted in: Schwartz, B. (1975). *Queuing and Waiting*, 43. Chicago: University of Chicago Press.

27. Mann, L. (1970). The social psychology of waiting lines. *American Scientist* 58, 389–98.

28. Milgram, S., Liberty, H., Toledo, R., and Wackenhut, J. (1986).

Response to intrusion into waiting lines. *Journal of Personality and Social Psychology* 51, 683–89.

29. Mann, L. (1970). The social psychology of waiting lines. *American Scientist* 58, 389–98.

30. Schwartz, B. (1975). *Queuing and Waiting*, 153–66. Chicago: University of Chicago Press

31. *The New Yorker* (1984, July 9). King Hassan of Morocco, 47.

CHAPTER SIX
WHERE IS LIFE FASTEST?

1. Reingold, E. M. (1987, February 2). A homecoming lament. *Time*, 55.

2. Overall scores were calculated by statistically standardizing the raw times for each experiment, so that the results from each experiment were weighted equally, and then adding up these three standardized scores.

3. More accurately, these are "non-ex-Soviet bloc" countries of Western Europe.

4. The perils of 1997. *Time* (1991, May 13th), 14.

5. Mean Streets. The New York Times, December 10, 1996, p. A13.

6. Whyte, W. (1988). *City*, 60. New York: Doubleday.

7. Whyte, W. (1988). *City*, 60–61. New York: Doubleday.

8. Riding, A. (1991, July 7). Why la dolce vita is easy for Europeans. *New York Times: The Week in Review*, 2; and: Sanger, D. (1991, July 7). As Japanese work ever harder to relax. *New York Times: The Week in Review*, 2.

9. JETRO (1992). *Nippon 1992 Business Facts & Figures*. Tokyo: Japan External Trade Organization.

10. Gasparini, G. (1995). On waiting. *Time and Society* 4, 29–45.

11. Hunnicutt, Benjamin J. (1988). *Work Without End*. Philadelphia: Temple University Press.

12. Hunnicutt, Benjamin J. (1996). *Kellogg's Six-Hour Day*. Philadelphia: Temple University Press, p.17.

13. Ibid., p. 35.

14. Ibid., p.145.

15. "France's Privileged Workers." Report on National Public Radio's Morning Edition, January 22, 1997.

16. Schor, Juliet B. (1991). *The Overworked American: The Unexpected Decline of Leisure*. New York: Basic Books.

17. European Trade Union Institute, *Collective Bargaining in Western Europe in 1988 and Prospects for 1989.* Brussels: EuroInt 1988/89, 62.

18. Hadenius, S., and Lindgren, A. (1992). *On Sweden.* Helsingborg, Sweden: The Swedish Institute.

19. Gibbs, N. (1989, April 24). *Time,* 58–67.

20. Quoted in: Rifkin, J. (1987). *Time Wars,* 51. New York: Henry Holt.

21. Pratt, L. (1981). Business temporal norms and bereavement behavior. *American Sociological Review* 46, 317–33.

22. McCaffery, R. (1972). *Managing the Employees Benefit Program,* 125. New York: American Management Association.

23. JETRO (1992). *Nippon 1992 Business Facts and Figures.* Tokyo: Japan External Trade Organization.

24. From: Respondents, E. H. & Respondents, P. K. (eds.) (1991). *Index to International Public Opinion, 1989–1990.* New York: Greenwood Press.

25. Godbey, G., and Graefe, A. (1993, April). Rapid growth in rushin' Americans. *American Demographics,* 26–28.

26. Robinson, J. P., and Godbey, G. (1996, June). The great American slowdown. *American Demographics,* 42–48.

27. Recent data from John Robinson and Ann Bostrum indicate that the number of hours that Americans recall working in retrospective surveys tends to be significantly greater than the amount they record in minute-to-minute diaries. This difference has grown from an average of one hour a week in 1965 to 7 hours in 1985. The memory gap is greatest among people who recall working very long hours. Those who estimate having worked between 50 and 54 hours average 9 fewer hours in their time diaries. Those who recall working 75 or more hours overestimate by an average of 25 hours. Similar comparisons are not available for other countries. All of the work and leisure hour data reported in this chapter are based on retrospective surveys. See: Russell, C. (1995, March). Overworked? Overwhelmed? *American Demographics,* 8, 51.

28. From: Respondents, E. H., and Respondents, P. K. (eds.) (1993). *Index to International Public Opinion, 1991–1992.* New York: Greenwood Press.

29. From: Respondents, E. H., and Respondents, P.K. (Eds.) (1991). *Index to International Public Opinion, 1989–1990.* New York: Greenwood Press.

30. Morris, J. (1984). "Trans-Texas." In *Journeys,* 111. Oxford: Oxford University Press.

31. This study was originally published as: Levine, R., Lynch, K., Miyake, K., and Lucia, M. (1989). The type A city: Coronary heart disease and the pace of life, *Journal of Behavioral Medicine* 12, 509–24.

32. City size was based on population estimates for the greater metropolitan area (what the Census Bureau currently refers to as a Metropolitan Statistical Area [MSA] or, in the case of some larger densely populated areas, a Primary Metropolitan Statistical Area [PMSA]) for each city. For further definition, see: U. S. Bureau of the Census (1991). *State and Metropolitan Area Data Book, 1991.* Washington, D.C.: U.S. Government Printing Office.

33. The elapsed times in the two situations were later statistically combined.

34. As in the earlier experiment, overall pace-of-life scores were derived by statistically standardizing the raw times for each experiment and then adding up these standardized scores.

35. *Los Angeles Times* (1989, October 22). The wristwatch factor, M4.

36. We had initially intended to include the proportion of individuals wearing watches in our first international study, but were undone by my lack of sensitivity to fashion trends. In Taiwan, in particular, I was at first surprised to find that almost no women seemed to be wearing watches. After returning home, however, I learned that I'd been looking in the wrong place. The current style among women, I was informed, was to wear watches around their necks, a fashion that had completely escaped me. As a result, these data had to be discarded.

CHAPTER SEVEN
HEALTH, WEALTH, HAPPINESS, AND CHARITY

1. Friedman, M., and Ulmer, D. (1984). *Treating Type A behavior.* New York: Random House.

2. Several researchers have been unable to reproduce the results found by Friedman and Rosenman. Many of them believe that the relationship of the Type A behavior pattern to coronary heart disease results from the toxicity of a single Type A component, what is called "hostility-anger." Time urgency, these studies argue, may also be characteristic of people high in hostility-anger, but it does not by itself lead to CHD. But other respected researchers in the Type A debate have found that time urgency, along with the related tendency to remain chronically active and "keyed up," may indeed lead to heart dis-

ease. (See, for example: Wright, L. [1988]. The Type A behavior pattern and coronary artery disease. *American Psychologist* 43 [1], 2–14). For reviews of the Type A literature, see: (1) Booth-Kewley, S., and Friedman, H. (1987). Psychological predictors of heart disease: a quantitative review. *Psychological Bulletin* 101, 343–62; and (2) Matthews, K. (1988). Coronary heart disease and Type A behaviors: Update on an alternative to the Booth-Kewley and Friedman (1987) quantitative review. *Psychological Bulletin* 104, 373–80. I return to this question in the next chapter.

3. Deaths from ischemic heart disease (a decreased flow of blood from the heart).

4. The correlation between overall pace of life and CHD death rates was .51 for the 36 U.S. cities and .35 for the 31 country study. Since age is positively correlated with the incidence of heart disease, we statistically adjusted the death rates for age—by taking into account, first, the median age and, second, the percentage of residents aged 65 and older—in each city. These adjustments did not significantly affect the magnitude of the correlations.

5. Smith, T., & Anderson, N. (1986). Models of personality and disease: An interactional approach to Type A behavior and cardiovascular risk. *Journal of Personality and Social Psychology* 50, 1166–73.

6. U.S. Department of Health and Human Services (1987). Regional variation in smoking prevalence and cessation: Behavioral risk factor surveillance, 1986. *Morbidity and Mortality Weekly Report* 36, 751–54.

7. Diener, E., Diener, M., and Diener, C. Factors predicting the subjective well-being of nations. Unpublished manuscript, University of Illinois, 1994.

8. To assess life satisfaction, we drew upon data from some of the large-scale surveys that are regularly conducted by governments and research groups in many countries. Survey data were available for about half (15) of the countries in our study. Life satisfaction ratings for each country were based on the last national survey taken in each country, as summarized by: Veenhoven, R. (1993). *Happiness in Nations.* Rotterdam: Risbo.

9. Schor, J. B. (1991). *The Overworked American,* 10. New York: Basic Books.

10. Diener, E. (March 23, 1995). Personal communication.

11. It should be noted that our data concern averages within countries. They do not tell us whether the same people in time-urgent cultures who are prone to heart attacks are also more likely to be

satisfied with other aspects of their lives, or whether there is simply a tendency for some people to thrive on a fast pace while a coexisting group suffers from it.

12. Raybeck, D. (1992). The coconut-shell clock: time and cultural identity. *Time and Society* 1 (3), 323–40.

13. Milgram, S. (1970). The experience of living in cities. *Science* 167, 1461–68.

14. With a single exception: Santa Barbara, California, was substituted for Oxnard, California.

15. The canes, and training for the role, were provided by the Fresno Friendship Center for the Blind.

16. The United Way contributions were for the year 1990, which was the latest available data at the time of the study. All of the field experiments were conducted in main downtown areas on clear summer days during primary business hours. A relatively equal number of male and female pedestrians were targeted. In all, we conducted 379 trials of the blind person episode, approached approximately 700 people in each of the dropped pen, hurt leg, and asking for change episodes, and left a total of 1,032 "lost" letters.

17. For the complete results of our U.S.A. helping experiments, see: Levine, R. (1993, October). Cities with heart. *American Demographics*, 46–54; and: Levine, R., Martinez, T., Brase, G., and Sorenson, K. Helping in 36 U.S. cities (1994). *Journal of Personality and Social Psychology* 67, 69–81.

18. As was the case for the pace data, there was generally not too much difference from one rank to the next. But at the extremes, the differences were again dramatic. In the dropped pen situation, for example, a stranger would have lost more than three times as many pens in Chicago before receiving help than they would have in Springfield, Massachusetts. When it came to making change for a quarter, nearly 80 percent checked their pockets in first-place Louisville, compared to 11 percent in last-place Patterson. My own home town of Fresno was dead last on two of the measures. We returned barely half as many (53 percent) letters as did San Diegans (100 percent!).

19. These experiments are part of a larger ongoing study on international differences in helping behavior. For more detailed description of these experiments, see: Levine, Martinez, Brase, and Sorenson (1994). Helping in 36 U.S. cities. *Journal of Personality and Social Psychology* 67, 69–81. For a preliminary report on the results of these studies, which are still in progress, see: Norenzayan, A., and Levine, R. (1994, April). Helping in 18 international cities. Paper

presented at the meetings of the Western Psychological Association, Kona, Hawaii.

CHAPTER EIGHT
JAPAN'S CONTRADICTION

1. *Business Week*, March 24, 1986.

2. Personal communication, July 15, 1987.

3. Vacation Competently. *Asahi Evening News* (1987, July 16).

4. It's official! vacations really aren't un-Japanese. *The New York Times International Edition* (1988, August 6), 4.

5. Sanger, D. (1991, July 7). . . . As Japanese work even harder to relax. *New York Times,* 2.

6. Ibid.

7. Quoted in: Sanger, D. E. (1993, May 30). The career and the kimono. *The New York Times Magazine,* 29.

8. Reynolds, G. (1995, April 27). E-Mail posting on the *Intercultural Network.*

9. Japan in the mind of America / America in the mind of Japan. *Time* (1992, February 10), 22.

10. *World health statistics annual, 1980: Vital statistics and causes of death.* Geneva: World Health Organization.

11. Marmot, M. G., and Syme, S. (1976). Acculturation and coronary heart disease in Japanese-Americans. *American Journal of Epidemiology* 104, 225–47.

12. Mishima, Y. *Sun and Steel.* Quoted in Stokes, H. (1975). *The Life and Death of Yukio Mishima,* 160. Tokyo: Charles E. Tuttle.

13. Reported in: Weisz, J. R., Rothbaum, F. M., and Blackburn, T. C. (1984). Standing out and standing in: The psychology of control in America and Japan. *American Psychologist* 39, 955–69.

14. Christopher, R. C. (1983). *The Japanese Mind: The Goliath Explained,* 70. Tokyo: Charles E. Tuttle.

15. Ibid., 148.

16. For a review of some of these studies, see: Taylor, S. E. (1991). *Health Psychology (2nd ed.).* New York: McGraw-Hill.

17. Williams, R. (1992). Prognostic importance of social and economic resources among medically treated patients with angiographically documented coronary artery disease. *Journal of the American Medical Association* 267, 520–24.

18. Cohen, J. B., Syme, S. L., Jenkins, C. D., Kagan, A., and Zyzan-

ski, S. J. (1975). The cultural context of Type A behavior and the risk of CHD. *American Journal of Epidemiology* 102, 434.

19. Appels, A. (1979). A psychosocial model of the pathogenesis of coronary heart disease. *Gedrag Ujdschrift Voor Psychologie*, 6–21.

20. Our findings are consistent with the assertion of many current Type A researchers that the single toxic Type A component is "hostility-anger." They argue that time urgency is a common characteristic of people high in hostility-anger—at least it is so in the United States—but it does not in itself lead to CHD.

21. Personal communication, July 22, 1987.

22. Sato, ibid.

23. Ibid.

24. Lawyer laments death by overwork. *The Japan Times Weekly Overseas Edition* (October 7, 1989), 5

25. Iyer, P. (1991). *The Lady and the Monk.* New York: Alfred Knopf.

26. Wood, C. (1986, September). The hostile heart. *Psychology Today*, 10–12, 12.

CHAPTER NINE
TIME LITERACY: LEARNING THE SILENT LANGUAGE

1. Mumford, L. (1963). *Technics and Civilization*, 18. New York: Harcourt, Brace, and World.

2. Rifkin, J. *Time Wars*, p.166.

3. Banfield, E. C. (1968). *The Unheavenly City: The Nature and Future of Our Urban Crisis*, 125–26. Boston: Little Brown.

4. Time is not on their side. *Time* (1989, February 27), 74.

5. Norton, D. G. (1990, April). Understanding the early experience of black children in high risk environments: Culturally and ecologically relevant research as a guide to support for families. *Zero to Three* 10, 1–7; and: Norton, D. G. (1993). Diversity, early socialization, and temporal development: The dual perspective revisited. *Social Work* 38, 82–90.

6. Lopez, V. Personal communication, June 6, 1995.

7. The complete program of activities is described in: Melitz, Z., Ben-Baruch, E., Hendelmen, S., and Friedman, L. (1993). *Time in the world of kindergarten children (experimental edition)*. Beersheva: University of Ben-Gurion in the Negev. (Materials are available from either of the first two authors).

8. Many of these categories are adapted from a talk by Richard

Brislin, which was part of a Symposium on Time and Culture presented at the Annual Meeting of the Western Psychological Association, Kona, Hawaii, April, 1994.

9. This example is adapted from: Brislin, R., Cushner, K., Cherrie, C., and Yong, M. (1986). *Intercultural Interactions: A Practical Guide.* Beverly Hills: Sage. Many other examples of critical incidents are described in this book. Another good compendium is: Brislin, R., and Yoshida, T. (1994). *Intercultural Communication Training: An Introduction.* Thousand Oaks, Calif.: Sage.

10. Brislin, Symposium on Time and Culture.

11. Ibid.

12. Hall, E. T. (1983). *The Dance of Life.* New York: Anchor Press.

13. Quoted in: Jackson, P., and Delehanty, H. (1995). *Sacred Hoops,* 169. New York: Hyperion.

14. Ibid.

15. Levine, R., Sato., S., Hashimoto, T., and Verma, J. (1995). Love and marriage in eleven cultures. *Journal of Cross-Cultural Psychology* 26, 554–71.

16. Bourguignon, E., and Greenbaum, L. (1973). *Diversity and Homogeneity in World Societies.* New Haven: HRAF Press.

17. Ibid.

18. Trifonovich, G. (1977). On cross-cultural orientation techniques. In R. Brislin (ed.), *Culture Learning: Concepts, Applications and Research,* 213–22. Honolulu: University Press of Hawaii.

19. Brislin, R., and Yoshida, T. (1994). *Intercultural Communication Training: An Introduction.* Thousand Oaks, Calif.: Sage.

20. These comments come from a personal interview on January 30, 1995. Altman repeated some of his comments in: Altman, N. (1995). *The Analyst in the Inner City,* 110–11. Hillsdale, N.J.: The Analytic Press.

21. Spengler, O. (1926). *The Decline of the West, Volume 1.* New York: Knopf.

CHAPTER TEN
MINDING YOUR TIME, TIMING YOUR MIND

1. This title is taken from: Michon, J. A. (1989). Timing your mind and minding your time. In Fraser, J. T. (ed.), *Time and Mind: Interdisciplinary Issues,* 17–39. Madison, Conn.: International Universities Press.

2. Storti, C. (1990). *The Art of Crossing Cultures*, 94–95. Yarmouth, Me.: Intercultural Press.

3. Pogrebin, L. C. (1996, May/June). Time is all there is. *Tikkun* 11, 43–47, 46.

4. See: Heschel, A. (1959). *Between God and Man: An Interpretation of Judaism.* New York: The Free Press. And: Heschel, A. (1960). The Sabbath—a day of armistice. In Greenberg, S., *A Modern Treasury of Jewish Thoughts,* 129. New York: Thomas Yoseloff.

5. Every forty-ninth year, according to some interpretations.

6. Except for houses in walled cities (unless bought of a Levite).

7. For further discussion of the Quiché sense of time, see: Tedlock, B. (1992). *Time and the Highland Maya (Revised Edition).* Albuquerque, N.M.: University of New Mexico Press. The Quiché are also discussed in: Hall, E. (1983). *The Dance of Life.* Garden City: Anchor Press.

8. Quoted in: Pogrebin, L. (1996, May/June). Time is all there is. *Tikkun* 11, 46.

9. Hoffman, E. (1989). *Lost in Translation,* 279. New York: Penguin.

10. Freedman, J., and Edwards, D. (1988). Time pressure, task performance, and enjoyment. In J. E. McGrath (ed.), *The Social Psychology of Time: New Perspectives,* 113–33. Newbury Park, Calif.: Sage.

11. Quoted in: Myers, D. (1992). *The Pursuit of Happiness,* 137. New York: Avon.

12. These and other related studies are described in ibid.

13. Greenberger, E., O'Neil, R., and Nagel, S. (1994). Linking workplace and homeplace: Relations between the nature of adults' work and their parenting behaviors. *Developmental Psychology* 30, 990–1002.

14. Levine, R., and Conover, L. (1992, July). The pace of life scale: Development of a measure of individual differences in the pace of life. Paper presented to the International Society for the Study of Time, Normandy, France; and: Soles, J. R., Eyssell, K., Norenzayan, A., and Levine, R. (1994, April). Personality correlates of the pace of life. Paper presented at the meetings of the Western Psychological Association, Kona, Hawaii.

15. McCrum, Robert (1996, May 27). My old and new lives. *The New Yorker,* 112–19, 118.

16. French, J. R., Jr., Caplan, R. D., and Harrison, W. (1982). *The Mechanisms of Job Stress and Strain.* New York: John Wiley; and: Caplan, R. D., Cobb, S., French, J. R., Jr., Harrison, R. V., and Pinneau, S. R.,

Jr. (1980). *Job Demands and Worker Health: Main Effects and Occupational Differences.* Ann Arbor, Mich.: Institute for Social Research.

17. Rosenman, R. (1987). The impact of anxiety and non-anxiety in cardiovascular disorders. Paper presented at a conference on Applications of Individual Differences in Stress and Health Psychology, Winnipeg, Manitoba, Canada.

18. Wright, L. (1988). The Type A behavior pattern and coronary artery disease. *American Psychologist* 43 (1), 2–14.

19. See, for example: Kelly, J. R. (1988). Entrainment in individual and group behavior. In McGrath, J. E. (ed.), *The Social Psychology of Time: New Perspectives,* 89–110. Newbury Park, Calif.: Sage; and: McGrath, J. (1989, June). The place of time in social psychology: Nine steps toward a social psychology of time. Paper presented at the meetings of the International Society for the Study of Time, Glacier Park, Montana.

20. Bem, S. L., Martyna, W., and Watson, C. (1976). Sex typing and androgyny: Further explorations of the expressive domain. *Journal of Personality and Social Psychology* 43, 1016–23.

21. Bradbury, T. N., and Fincham, F. D. (1988). Individual difference variables in close relationships: A contextual model of marriage as an integrative framework. *Journal of Personality and Social Psychology* 54, 713–21.

22. Quoted in: Schwartz, T. (1995). *What Really Matters,* 367. New York: Bantam.

23. Rifkin, J. (1987). *Time Wars.* New York: Henry Holt.

24. Quoted in Keyes, R. (1991). *Timelock,* 192. New York: Ballantine.

25. Quoted in: Govinda, Lama Anagarika. Introductory forward, lii-lxiv. In: Evans-Wentz, W. Y. (1960). *The Tibetan Book of the Dead,* lx. New York: Oxford University Press.

26. Pinker, S. (1994). *The Language Instinct,* 209. New York: William Morrow; quoted in an e-mail posting by Mark Aultman on ISSTL@PSUVM.PSU.EDU, February 26, 1996.

27. Itchy feet and pencils: A symposium (1991, August 18). *The New York Times Book Review,* 3, 23.

INDEX

Pace of life studies (*cont.*)
149–51, 158, 161; *La Dolce Vita* and, 139–46; physical well-being and, 154–57; psychological well-being and, 157–59, 212–13; slow pace in, 135–39, 164–65; social well-being and, 159–68; talking speeds in, 147–49, 151; in 36 cities, 146–49; in 31 countries, 131–46; walking speeds and, 8, 16–17, 20, 130–32, 147–51; work speeds, 9, 130–31, 147, 148–49
Pakistan, collectivism in, 18
Palmer, Willard, 3
Papua New Guinea, 14
Partin, Glenn, 117
Peace Corps volunteers, 5–6, 86, 202, 204–6
Pendulum clocks, 55, 57
Pepto-Bismol, 39–40
Person-environment (P-E) fit, 215–17
Phillips, Mark, 5
Physical well-being, 154–57
Plautus, 72
Pleasantness, 38–39
Pogrebin, Letty, 208
Poland, waiting time in, 106–7
Polychronic (P-time) schedulers, 96–97, 98, 201
Population size, 16–17
Post, Emily, 123–24, 144
Pratt, Lois, 144
Primitive agricultural societies, 14
Priority Management Pittsburgh Inc., 102
Psychological androgyny, 217–18
Psychological clock, 26–50; bore-

dom and, 36–37; bumps in time and, 48–50; discrepancy between people, 77–78; distortion of, 28–32; influences on, 37–48; stretching time and, 33–37
Psychological overload, 160
Psychological well-being, 157–59, 212–13
Punctuality: culture and, 193–96, 200–201; importance of, 59, 69–70; origins of concept, 57; success and, 110–11; virtues of, 67–68

Queue breaking, 124–25
Queuing and Waiting (Schwartz), 113, 114–15
Quich, of Guatemala, 210–11

Radecki, Sigmund von, 58
Railroads, 65–67, 73
Ranjeesh, Bhagwan Shree, 119
Raybeck, Douglas, 93–94
Reingold, Edwin, 130
Reis, Harry, 163
Remembered duration (Block), 29
Repent Harlequin (Ellison), 51
Revolutionary calendars, 78–79
Reynolds, Garr, 172–73
Riding, Alan, 139–40
Rifkin, Jeremy, xi, 70–71, 72, 75, 76, 79, 188, 189, 219
Rilke, Ranier Maria, 122
Ritter, Jean, 165
Rituals, 208–10
R-mode, 46–48
Robinson, John, 145
Rogers, Richard, 117
Romantic love, 199–200

Switzerland: climate and, 17; pace of life in, 132, 134
Syme, Leonard, 174
Synchronization, 63–67
Systems and Procedures Association of America, 71
Szalai, Alexander, 60

Tagaki, Ikuro, 171
Taj Mahal, 90
Talking speed, 147–49, 151
Taylor, Frederick, 70–72, 74
Taylorism, 70–72, 74
Television, 45
Tempo, 3–25; elements of, 8–19, 27; individual differences in, 19–24; moving beyond, 24–25; perceived versus real, 27–28; time signatures around the world, 5–8
Temporal Man (Lauer), 86
Temporal territoriality, 77
Tempostasis, 77
Tennis, 33
Thich Nhat Hanh, 41
Thoreau, Henry David, xii, 52, 214
Thutmose III, 54
Tibet, collectivism in, 18
Tibetan Book of the Dead, 223
Tiger at the Gates (Giraudoux), 101
Time-and-motion studies, 70–72, 74
"Time as money" attitude, 90–91, 101–7, 201
Time-free tasks, 45–48
Time literacy, 187–206; accepted sequences and, 198–200; clock time versus event time and, 200–201; criticism and, 202–3; "doing nothing," 197–98; flexibility in, 203–6; in other cultures, 190, 204–6; practicing, 202; punctuality and, 193–96, 200–201; teaching time in, 191–92; waiting game and, 197; work time versus social time and, 196
Time perception studies, 26–27, 32
Timepieces, 53–59; accuracy of, 131–32, 137, 147; early, 53–56; industrialization and, 62; mechanical, 56–59
Time pressure, 23–24, 212–13
Time recording systems, 68–69
Time-selling, 115–18
Time-stretching: boredom and, 36–37; examples of, 33–35
Time wars, 76–79
Time Wars (Rifkin), xi
Time zones, 65
Tiv of Nigeria, 14–15
Tocqueville, Alexis de, 207
Toffler, Alvin, 4
Tramites (bureaucratic procedures), 115
Traore, Jean, 91
Triandis, Harry, 18–19
Trifonovich, Greg, 202
Trinidad: colored-people's time (CPT) in, 86–87; time signature of, 7–8
Trobriand Islanders, 61
Turk, Fred, 44–45
Type A behavior pattern, xix, xviii, 15, 19–22, 92; coronary heart disease and, 155–57, 174, 177–78, 183, 216–17; in Japan, 170, 174, 177–78, 183; in the United States, 155–57

Type B behavior pattern, xviii, 155, 156, 216

Ueda, Keiko, 43
Ulmer, Diane, 21–22
Unheavenly City, The (Banfield), 188
United States, 48, 153–68;
amount of activity and, 41;
colored people's time (CPT)
in, 10–11, 86–87; economic
factors in, 10–11; importance
of change in, 44–45; individualism in, 18, 159; pace of life
in, 16–17, 130, 134–35, 136,
138; physical well-being in,
154–57; psychological well-being in, 157–59; social well-being in, 159–68; time-selling in,
117–18; vacation time in, 144;
waiting time in, 102; workaholism in, 139–40, 141
University of Michigan, 216, 219
Urgency, 21–23, 39–40, 158,
178

Vacation time, 143–44, 145,
171–72
Vanderbilt, Amy, 144
Variety, 44–45
Verbal communication, 41–44;
in Japan, 42–43; silence in,
41–42; speaking order in,
43–44; speech patterns in, 20
Voices (Kir-Stimon), 77–78
Vonnegut, Kurt, 74

Waiting, 21, 101–26; ability to
control time and, 118–23,
125–26; appointment times
and, xiv-xv, 87–88, 110–11,
193–96; international forums
and, 119–21, 125–26; military
strategy and, 119–21; money
and privilege in, 114–18; pain
caused by, 101–2; punctuality
and, 57, 59, 67–68, 69–70,
110–11, 193–96, 200–201;
queue breaking and, 124–25;
Siddhartha move and,
121–23, 197; status and, xiv-
xv, 109–14, 118; supply and
demand in, 106–7; time as gift
and, 123–24; time as money
and, 90–91, 101–7, 201; time
literacy and, 197; value and,
107–9
Waiting for Godot (Beckett), 37,
113
Walden (Thoreau), 52
Waldo, Leonard, 65, 66–67
Walking speeds, 8, 16–17, 20,
130–32, 147–51
Warner, Charles Dudley, 75
Wasting of time, 91, 117–18,
197–98
Waterbury Watch Company, 69
Water clocks, 54–55, 56
Watts, Alan, 33
Weather Bureau, 65
Weight-driven timepieces, 56–57
Weissman, Debbie, 211
Well-being: physical, 154–57;
psychological, 157–59,
212–13; social, 159–68
West, Nathanael, 52
Western Collaborative Group
Study, 155
Western Union, 66
West Germany, workaholism in,
140
White, E. B., 111–12